RNA-seq in Drug Discovery and Development

The research and development process in modern drug discovery and development is a complex and challenging task. Using traditional biological test methods such as PCR to measure the expression levels or function of these genes is costly and time-consuming. RNA-seq can measure the expression patterns of thousands of genes simultaneously and provide insights into functional pathways or regulations in biological processes, which has revolutionized the way biological scientists examine gene functions. This book addresses the various aspects of the RNA-seq technique, especially its application in drug discovery and development.

Features

- One of the few books that focuses on the applications of the RNA-seq technique in drug discovery and development
- Comprehensive and timely publication which relates RNA sequencing to drug targets, mechanisms of action, and resistance
- The editor has extensive experience in the field of computational medicinal chemistry, computational biophysics, and bioinformatics
- Chapter authors are at the frontline of the academic and industrial science in this particular area of RNA sequencing

Drugs and the Pharmaceutical Sciences
A Series of Textbooks and Monographs

Series Editor
Anthony J. Hickey
RTI International, Research Triangle Park, USA

The Drugs and Pharmaceutical Sciences series is designed to enable the pharmaceutical scientist to stay abreast of the changing trends, advances, and innovations associated with therapeutic drugs and that area of expertise and interest that has come to be known as the pharmaceutical sciences. The body of knowledge that those working in the pharmaceutical environment have to work with, and master, has been expanding, and continues to expand, at a rapid pace as new scientific approaches, technologies, instrumentations, clinical advances, economic factors, and social needs arise and influence the discovery, development, manufacture, commercialization, and clinical use of new agents and devices.

RECENT TITLES IN SERIES

Good Design Practices for GMP Pharmaceutical Facilities,
Second Edition
Terry Jacobs and Andrew A. Signore

Handbook of Bioequivalence Testing, Second Edition
Sarfaraz K. Niazi

FDA Good Laboratory Practice Requirements, First Edition
Graham Bunn

Continuous Pharmaceutical Processing and Process Analytical Technology
Ajit Narang and Atul Dubey

Project Management for Drug Developers
Joseph P. Stalder

Emerging Drug Delivery and Biomedical Engineering Technologies:
Transforming Therapy
Dimitrios Lamprou

RNA-seq in Drug Discovery and Development
Feng Cheng and Robert Morris

For more information about this series, please visit www.crcpress.com/Drugs-and-the-Pharmaceutical-Sciences/book-series/IHCDRUPHASCI

RNA-seq in Drug Discovery and Development

Edited By

Feng Cheng
Associate Professor, Department of Pharmaceutical Sciences,
Taneja College of Pharmacy, University of South Florida

Robert Morris
Department of Pharmaceutical Science,
Taneja College of Pharmacy, University of South Florida

CRC Press is an imprint of the
Taylor & Francis Group, an **informa** business

First edition published 2024
by CRC Press
6000 Broken Sound Parkway NW, Suite 300, Boca Raton, FL 33487-2742

and by CRC Press
4 Park Square, Milton Park, Abingdon, Oxon, OX14 4RN

© 2024 selection and editorial matter, Feng Cheng and Robert Morris; individual chapters, the contributors

CRC Press is an imprint of Taylor & Francis Group, LLC

Reasonable efforts have been made to publish reliable data and information, but the author and publisher cannot assume responsibility for the validity of all materials or the consequences of their use. The authors and publishers have attempted to trace the copyright holders of all material reproduced in this publication and apologize to copyright holders if permission to publish in this form has not been obtained. If any copyright material has not been acknowledged please write and let us know so we may rectify in any future reprint.

Except as permitted under U.S. Copyright Law, no part of this book may be reprinted, reproduced, transmitted, or utilized in any form by any electronic, mechanical, or other means, now known or hereafter invented, including photocopying, microfilming, and recording, or in any information storage or retrieval system, without written permission from the publishers.

For permission to photocopy or use material electronically from this work, access www.copyright.com or contact the Copyright Clearance Center, Inc. (CCC), 222 Rosewood Drive, Danvers, MA 01923, 978-750-8400. For works that are not available on CCC please contact mpkbookspermissions@tandf.co.uk

Trademark notice: Product or corporate names may be trademarks or registered trademarks and are used only for identification and explanation without intent to infringe.

Library of Congress Cataloging-in-Publication Data
Names: Cheng, Feng (Computational biologist), editor. |
Morris, Robert (Of University of South Florida), editor.
Title: RNA-seq in drug discovery and development / edited by Feng Cheng, Robert Morris.
Other titles: Drugs and the pharmaceutical sciences. 0360-2583
Description: First edition. | Boca Raton : CRC Press, 2023. |
Series: Drugs and the pharmaceutical sciences series |
Includes bibliographical references and index. |
Identifiers: LCCN 2023004854 (print) | LCCN 2023004855 (ebook) |
ISBN 9781032004068 (hardback) | ISBN 9781032004099 (paperback) |
ISBN 9781003174028 (ebook)
Subjects: MESH: Sequence Analysis, RNA—methods | Drug Discovery—methods |
Computational Biology—methods
Classification: LCC QP623.5.S63 (print) | LCC QP623.5.S63 (ebook) |
NLM QU 550.5.S4 |
DDC 572.8/8—dc23/eng/20230503
LC record available at https://lccn.loc.gov/2023004854
LC ebook record available at https://lccn.loc.gov/2023004855

ISBN: 978-1-032-00406-8 (HB)
ISBN: 978-1-032-00409-9 (PB)
ISBN: 978-1-003-17402-8 (EB)

DOI: 10.1201/9781003174028

Typeset in Times
by codeMantra

Contents

Preface vii
Editors ix
Contributors xi

Chapter 1   Introduction to RNA Sequencing and Quality Control 1

Robert Morris and Feng Cheng

Chapter 2   Read Alignment and Transcriptome Assembly 35

Robert Morris and Feng Cheng

Chapter 3   Normalization and Downstream Analyses 61

Robert Morris and Feng Cheng

Chapter 4   Constitutive and Alternative Splicing Events 101

Robert Morris and Feng Cheng

Chapter 5   The Role of Transcriptomics in Identifying Fusion Genes
and Chimeric RNAs in Cancer 129

Robert Morris, Valeria Zuluaga, and Feng Cheng

Chapter 6   MiRNA and RNA-seq 153

Robert Morris and Feng Cheng

Chapter 7   Toxicogenomics and RNA-seq 171

Robert Morris, Kyle Eckhoff, Rebecca Polsky, and Feng Cheng

Chapter 8   Herbal Medicine and RNA-seq 205

Robert Morris and Feng Cheng

Chapter 9   Single-Cell RNA Sequencing 233

Robert Morris and Feng Cheng

Index 263

Preface

The advent of RNA-sequencing and transcriptomic technology in the 1970s has precipitated a greater understanding of molecular biology and the underlying mechanisms that drive pathology of disease and drug reactions. The evolution of sequencing technology, from Sanger sequencing to microarrays and modern high-throughput sequencing techniques, has vastly contributed to the current understanding of many facets of molecular biology, toxicology, cancer biology, and beyond. While there are many software and applications for transcriptomic technologies, this textbook will focus primarily on software available in the open-source platform Galaxy as well as R packages, which will also be accessed through the Galaxy interface and utilized for various downstream analyses. In addition, this textbook will provide examples of various applications for RNA-sequencing with a focus on drug discovery, drug interactions, and traditional medicines.

Our motivation in this book is to:

1. Provide an overview of the big picture about the RNA-seq technology including sequencing platforms, methodologies for sample preparation, quality control, alignment tools, and methods of downstream analysis.
2. Discuss software that are all open source and readily available for readers to learn step-by-step and perform the provided protocols. All software utilized for examples in this textbook include images and figures displaying the exact option parameters utilized.
3. Provide recent applications of RNA-seq on drug discovery. The hope is that readers can learn how to use RNA-seq and apply the foundational knowledge acquired from reading this textbook to solve novel sequencing-related problems.

EACH CHAPTER

This RNA-sequencing textbook will contain nine chapters pertaining to RNA-sequencing, drug discovery, and traditional medicine. Chapters 1–3 will provide an expansive overview of a traditional RNA-sequencing protocol. Specifically, Chapter 1 will introduce the concept and idea behind RNA-sequencing as well as introduce various methods of sample preparation. In addition, this chapter will discuss the advantages of RNA-sequencing over microarray technology and introduce a popular tool for sample quality control. Finally, Chapter 1 will introduce how the sequencing instrument works, the types of sequencing instruments commercially available, and basic information concerning the use of the public Galaxy interface. Chapter 2 will introduce multiple alignment tools available in Galaxy including Bowtie, cufflinks, and HISAT2 that can be used to align our raw sequencing reads produced from the sequencing instrument. Chapter 3 will

vii

discuss various normalization techniques that can be applied to aligned reads as well as discuss the primary forms of downstream analysis including differential expression, generation of co-expression networks, heatmaps, and volcano plots. In addition, examples using publicly available datasets will be presented for these tools in the Galaxy interface as well as step-by-step instructions on how to run these tools in Galaxy.

Chapters 4–9 will focus on the applications of these fundamentals of RNA-sequencing discussed in previous chapters. In Chapter 4, we will introduce alternative splicing and tools to identify splice junctions by either differences in exon usage or differential expression of individual isoforms across samples. Chapter 5 will expand upon splicing and introduce available RNA-sequencing tools to identify the presence of fusion genes and fusion transcripts. Chapter 6 will briefly introduce non-coding RNAs and a popular tool, miRDeep2, that can be used to assess the expression of miRNAs both within a single sample and across samples. Chapter 7 will introduce the role of RNA-sequencing techniques in toxicogenomics, while Chapter 8 will introduce how RNA-sequencing can be used to expand upon traditional medicine and identify pharmacologically relevant natural compounds. Finally, Chapter 9 will briefly introduce the emerging field of single-cell RNA-sequencing and describe this new approach to transcriptome analysis.

Editors

Dr. Cheng is a computational biologist. He received systematic training in bioinformatics at the Chinese Academy of Science, University of Illinois at Urbana Champaign, Rice University, University of Virginia, and Yale University. His research is mainly focused on applications of bioinformatics methods for the identification of drug–drug interactions (DDIs) and for drug repositioning by analyzing Big Data from public resources. He is also interested in computer-aided drug design. Dr. Cheng has published over 100 research papers in prestigious journals including *Nature, Neuron, Journal of the American Chemical Society*, and *Journal of Alzheimer's Disease*. His work has been cited more than 4,600 times. Presently, Dr. Cheng is the Editor-in-Chief of the *Open Bioinformatics Journal*. Dr. Cheng has been serving as an editorial member of several journals.

Mr. Morris is a computational biologist with an M.S in molecular biology and is currently working on his MPH in applied biostatistics. His research primarily is focused on using bioinformatic tools to identify drug-drug interactions as well as identify risk factors associated with elevated risk of adverse drug reactions including bradycardia and dysphagia using publicly available Big Data. He also is focused on elucidating drug mechanisms using bioinformatics and RNA sequencing techniques. He currently has 10 publications in a variety of journals including the *Journal of Alzheimer's Disease* and *Brain Sciences*. Finally, he has recently served as a peer reviewer for the *Drugs – Real World Outcomes* (DRWO) journal.

Contributors

Feng Cheng
Department of Pharmaceutical
 Sciences
Taneja College of Pharmacy,
 University of South Florida
Tampa, Florida

Kyle Eckhoff
Department of Cell Biology,
 Microbiology, and Molecular
 Biology
College of Art and Science, University
 of South Florida
Tampa, Florida

Robert Morris
Department of Pharmaceutical
 Science
Taneja College of Pharmacy,
 University of South Florida
Tampa, Florida

Rebecca Polsky
Biomedical Sciences Program
University of South Florida
Tampa, Florida

Valeria Zuluaga
Biomedical Sciences Program
University of South Florida
Tampa, Florida

1 Introduction to RNA Sequencing and Quality Control

Robert Morris and Feng Cheng
University of South Florida

CONTENTS

1.1 What Is RNA Sequencing? 2
    1.1.1 What Is RNA? 2
    1.1.2 What Is RNA-seq? 2
1.2 cDNA Library Preparation in RNA-seq 4
    1.2.1 Isolation of the RNA and RNA Quality Check 4
    1.2.2 Selection and Depletion of Particular RNA 6
    1.2.3 Fragmentation 6
    1.2.4 Reverse Transcription to Generate cDNA and Adaptor
            Sequences 6
    1.2.5 Single-End and Paired-End Sequencing Technique 8
1.3 RNA-seq Techniques and Platforms 9
    1.3.1 Roche 454 9
    1.3.2 Illumina Platform 10
    1.3.3 Small-Scale RNA-seq Platform 11
    1.3.4 Third-Generation Sequencing 11
1.4 RNA-seq File Format 12
    1.4.1 Output File: FASTQ File 12
    1.4.2 Mapped File: SAM/BAM/BIGWIG Formats 15
    1.4.3 GTF File 15
    1.4.4 BED File 17
1.5 Quality Control of RNA-seq Data 20
    1.5.1 Basic Usage of the Public Server Galaxy 20
    1.5.2 FastQC Program 20
    1.5.3 Trimmomatic Tool for Adapter Trimming 29
1.6 Advantages of RNA-seq Over Expression Microarrays 30
1.7 Summary 31
Keywords and Phrases 32
Bibliography 32

DOI: 10.1201/9781003174028-1

1

1.1 WHAT IS RNA SEQUENCING?

1.1.1 WHAT IS RNA?

Ribonucleic acid (RNA) is a very important biomolecule. A strand of RNA, which usually exists in a single-stranded form, has a backbone composed of alternating sugar units and phosphate groups. In addition, each sugar in the backbone of an RNA molecule is attached via a glycosidic bond to one of four nitrogenous bases: adenine (A), cytosine (C), guanine (G), or uracil (U).

Three distinct types of RNA are involved in the transcription and subsequent synthesis of proteins including messenger RNA (mRNA), transfer RNA (tRNA), and ribosomal RNA (rRNA). mRNAs are directly transcribed from genomic DNA during the molecular process of transcription and provide the blueprint needed for generation of proteins. tRNAs have a cloverleaf structure and are responsible for bringing amino acids to the site of translation. Finally, rRNAs compose the large and small subunits of ribosomes, of which are the molecular complexes required to connect mRNA and tRNA and synthesize new proteins. In addition, various non-coding RNAs (ncRNAs) exist within the cell and play many regulatory roles with regard to proper cell function and gene expression.

A transcriptome encompasses the entirety of RNA content in a cell including coding mRNA and ncRNA species including microRNAs, long-noncoding RNAs, and long intergenic non-coding RNAs. The quantification of RNA in a biological sample gives insight into transcription regulation, gene expression, temporal activation and deactivation of gene activity, alternative splicing events, and exon/intron boundaries.

1.1.2 WHAT IS RNA-SEQ?

Generally, RNA sequencing (RNA-seq) technology refers to any high-throughput approach to sequencing samples of RNA in massive parallel, that is to sequence the entirety of the transcriptome at once. RNA-seq can generate a snapshot of the current transcriptomic profile of a tissue sample, a single cell, or a group of cells. We can identify genes that are highly expressed or largely silenced in expression in cells at the point of analysis by mapping sequences to existing transcripts. In addition, RNA-seq techniques can identify single-nucleotide polymorphisms (SNPs).

In many scenarios, we will be concerned with the change in the transcriptomic profile between a control group of cells and an experimental group of cells. For instance, we may want to determine alterations in gene expression in a tissue sample that was exposed to an environmental toxin such as asbestos or mercury. As a result, we can identify differentially expressed genes, that is, genes that are transcribed at different rates between cells and/or conditions, and identify which genes may be responsible for regulation of a particular cellular response. Figure 1.1 provides an overview of the standard RNA sequencing pipeline protocol. In the case of a two-condition experimental design, we then lyse our cells

Introduction to RNA Sequencing and Quality Control

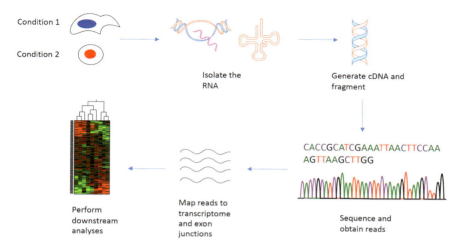

FIGURE 1.1 An overview of the RNA-seq process. In most RNA-seq workflows, we start with isolating RNA species from our samples of interest, which is usually comparing the transcriptomic changes between a control group and an experimental group. Once the RNAs are isolated, we convert them into cDNA to improve stability and decrease degradation through the use of platform-specific adapter sequences that are ligated to the 3' and 5' ends. The amplified cDNA is then fragmented into small pieces, typically 50–200 bp, in order to generate a large library of cDNA. The cDNA library is sequenced by a high-throughput RNA-seq platform. The raw reads data generated from the sequencing process then go through several filtering and normalization steps prior to performing reads alignment and downstream analyses such as differential expression analysis, co-expression networks, and isoform discovery.

and then isolate the RNA from the remaining cellular content. The RNA is then converted into complementary DNA (cDNA) to improve stability and then is fragmented into smaller pieces, the size of which is dependent on the constraints of the sequencing platform that is being used. We then sequence our samples, generating FASTA files that can then be subsequently refined and filtered using the tool FastQC to generate FASTQ files. These raw reads can then be aligned to a reference genome or transcriptome to generate binary alignment map (BAM)/ sequence alignment map (SAM) files containing mapping coordinates, which can then be used to perform various downstream analyses.

The gene/transcript expression level can be evaluated by counting the mapped reads in the SAM/BAM files. The expression levels are then normalized to remove, at least partially, intrasample variance among samples. RPKM, otherwise known as reads per kilobase per million mapped reads, is a normalization technique that accounts for the length of a transcript and is used in RNA sequencing protocols. Although the RPKM will frequently be reported by the sequencing software in the generated raw reads data, it can be calculated by using the following formula (Eq. 1.1):

$$\text{Gene RPKM} = \left(\frac{\text{\# of mapped reads to a gene} \times 10^3 \times 10^6}{\text{Total \# of mapped reads} \times \text{gene length in base pairs}} \right) \quad (1.1)$$

where 10^3 normalizes for gene length and 10^6 accounts for sequencing depth. For paired-end reads, we can instead calculate the analogous fragments per kilobase per million (FPKM) which takes into account that more than 1 read may correspond to the same fragment. More details of sample normalization techniques will be described in Chapter 3.

While there are multiple protocols to follow and multiple programs for RNA-seq analysis, this book will focus on the Illumina sequencing protocol as well as how to use the various tools in the Galaxy platform for each step of the protocol.

1.2 cDNA LIBRARY PREPARATION IN RNA-seq

Typically, the four steps for cDNA library preparation are as follows:

1. Isolation of the RNA and RNA quality check
2. Selection and depletion of particular RNA
3. Fragmentation
4. Reverse transcription to generate cDNA

1.2.1 ISOLATION OF THE RNA AND RNA QUALITY CHECK

As we are only concerned with the RNA population for transcriptome studies, the RNA from each sample needs to be separated from the genomic DNA still present. This process is done by incorporating DNase into the tissue samples, ultimately degrading most of the genomic DNA and leaving behind the RNA population of interest. Although any DNase could theoretically work, different enzymes may be more optimal than others depending on the sequencing platform utilized. In order to generate an RNA-seq library compatible with Illumina sequencing technology, the RNeasy mini kit from QIAGEN is frequently used to maximize the quality and yield of eukaryotic RNA. However, there are a variety of different preparation kits that can be utilized depending on the species origin of the tissue samples.

Due to the relative instability of RNA, isolation of RNA is immediately followed by either capillary electrophoresis or gel electrophoresis in order to quantify the degree of degradation and overall concentration of the sample. The sample is then assigned an RNA integrity (RIN) value that ranges from 1 to 10 for mammalian-derived samples with samples containing a lower degree of degradation being assigned a higher value on the scale.

Traditional methods of measuring RNA degradation, largely in part due to its low cost and ease of analysis, relied upon gel electrophoresis coupled with ethidium bromide staining. In this method, the ratio of the 28S peak, which corresponds to the large ribosomal subunit, and the 18S peak, which corresponds to

Introduction to RNA Sequencing and Quality Control 5

rRNA of the small ribosomal subunit, is estimated with a ratio of at least 2:1 indicating a high-quality sample. Because 28S rRNA degrades at a faster rate than 18S, a higher concentration of 28S rRNA is indicative of a higher quality sample.

Despite this, several biases and errors are associated with this method including visualization inconsistencies, loading errors, and complications when sample concentration is low. To rectify these inconsistencies and produce a new standardized approach, a Bayesian-based algorithm was constructed by Agilent Technologies by analyzing hundreds of RNA-seq samples and assigning a value from 1 to 10, thus generating a model that can be utilized to predict the degree of degradation for additional RNA-seq samples.

Generation of the RIN through the computational method differs somewhat from the degree of RNA integrity estimated by traditional methods. One such distinction is that the total RNA ratio is determined by taking the ratio of the areas underneath the 18S and 28S peaks and comparing it to the total area under the graph. In this method, the area underneath the 28S peak should be quite large and the area ratio value of the region to the left of the 18S peak, referred to as the fast region, should be small. Figure 1.2 displays a typical representation of a high-quality sample (RIN = 10), a fair-quality sample (RIN = 5), and a poor-quality sample (RIN = 1). For most RNA-seq experiments, an RIN value of 8 or greater is preferred.

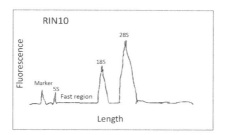

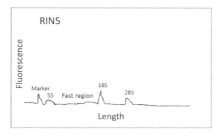
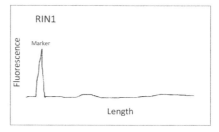

FIGURE 1.2 Utilizing the Bayesian-based algorithm designed by Agilent Technologies, the RIN for a sample can be calculated. On this scale, a value ranging from 1 to 10 is assigned to each sample with ten representing a high-quality sample with little degradation and 1 representing a low-quality sample with a high degree of RNA degradation. As 28S rRNA degrades at a faster rate, a lower 28S peak is indicative of a greater degree of decay in the sample. As you shift from a 10 to a 1 on the scale, the 28S and 18S peaks decrease significantly and the overall length of fragments declines.

1.2.2 Selection and Depletion of Particular RNA

Many RNA-seq studies will want to focus on a particular class or segment of the total RNA population in what is referred to as RNA enrichment or RNA selection. Because approximately 95% of the total RNA is rRNA, which is typically of low experimental relevance, it is important to eliminate this significant form of background noise prior to downstream analysis. An additional source of background noise that can hinder downstream analysis is the presence of immature, unprocessed RNA known as pre-mRNA.

There are three primary enrichment strategies for RNA selection depending on the type of RNA that is desired for analysis. The first method is PolyA selection which is performed in order to isolate mature mRNA from the total RNA sample. This is done by mixing the total RNA sample with poly(dT) oligomers, which are attached to a substrate such as magnetic beads, that will bind to and isolate mRNAs while excluding ncRNA species and unprocessed mRNA. As a result, some transcriptome representation will be lost as the expression of ncRNAs will be underestimated. A second approach is to deplete the population of rRNA itself, leaving behind transcripts that correspond to both non-coding and coding RNAs. The final approach is to design a capture method that will selectively isolate transcripts of interest at the cost of excluding large portions of the transcriptome. Thus, the enrichment strategy that should be performed is dependent on the parameters of the experiment and the research questions. Figure 1.3 summarizes these three enrichment methods as well as their most significant characteristics.

1.2.3 Fragmentation

Once the population of RNA has been either depleted or selected for the species of interest, RNA samples will then undergo a fragmentation process prior to sequencing. Longer RNA transcripts have to be segmented into smaller fragments, typically no larger than 500–600 nucleotides, due to the requirements of the sequencing instruments. The exact length required for the RNA sample is dependent on both the instrument model and the platform that is being used for the sequencing experiment. RNA strands can be fragmented using either various alkaline solutions, solutions containing high concentrations of Mg or Zn ions, or through the use of enzymes such as RNase III. Alternatively, RNA strands can first be reverse transcribed into full-length cDNA and then subsequently lysed into smaller fragments using DNases.

1.2.4 Reverse Transcription to Generate cDNA and Adaptor Sequences

One of the final steps of library preparation is conversion of the isolated transcripts into cDNA through use of reverse qRT-PCR. There are three reasons for performing this conversion:

Introduction to RNA Sequencing and Quality Control

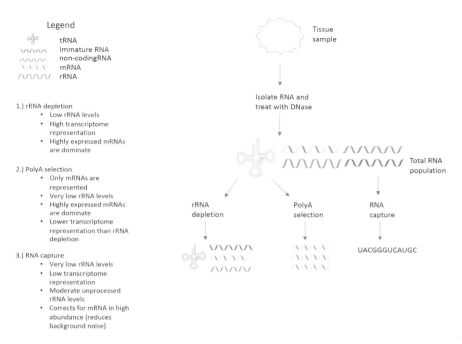

FIGURE 1.3 Three primary enrichment strategies that can be used depending on the RNA species of interest. The first enrichment strategy is rRNA depletion which results in low rRNA levels and high transcriptome representation as both coding and non-coding RNA will be present. The second method, PolyA selection, isolates exclusively mature mRNAs and is characterized by a somewhat lower degree of transcriptome representation than rRNA depletion. The final method is RNA capture which enables isolation of transcripts containing a particular sequence of interest. This results in an RNA library with low transcriptome representation and moderate levels of unprocessed RNA transcripts.

1. RNA is highly unstable outside of a cellular environment. To ensure a high-quality sample and minimize transcriptome loss, RNA is first converted to cDNA.
2. In many instances, the overall RNA sample will be quite small. In order to amplify the sample and increase the concentration of lowly expressed transcripts, polymerase chain reaction (PCR) is performed, which requires DNA as a template for DNA polymerase activity.
3. Sequencing instruments primarily require a DNA input.

Once the cDNA library has been generated, we will then ligate adaptor sequences to our cDNA strands. Adaptor sequences are small oligonucleotides that are incorporated into the ends of cDNA. Adaptors are necessary for PCR amplification of our samples so that they can be detected using the sequencing instrument as well as they contain barcoding sequences and binding sequences that are required for binding to the flow cell of the sequencing instrument.

1.2.5 Single-End and Paired-End Sequencing Technique

A single-end read is generated when the sequencing instrument reads a fragment starting from one end. In contrast, a paired-end read is generated when a fragment is read from both a forward direction and a reverse direction, resulting in match pairs for each sequence that contains reads for both sense and antisense template strand orientation (Figure 1.4). Single-end technique would be half as expensive as paired-end sequencing, assuming an equivalent number of sequenced cDNA fragments. Paired-end reads are preferable when the research question of interest is concerned with novel transcript discovery, isoform expression analysis, or for the annotation of a poorly characterized genome.

Paired-end reads, albeit more expensive to generate than single-end reads, are advantageous for multiple reasons. First, they generate reads with longer read lengths (longer contigs), allowing for more accurate genome assembly. If we refer to the puzzle analogy, paired-end reads generate larger puzzle pieces and allow for alignment tools to connect areas of the genome with greater certainty. In particular, paired-end reads are useful for identifying various genomic rearrangements and mutations including insertions, deletions, novel transcripts, and gene fusions, many of which cannot be correctly identified using single-end reads. Next, genomic regions containing long repetitive sequences are easier to correctly

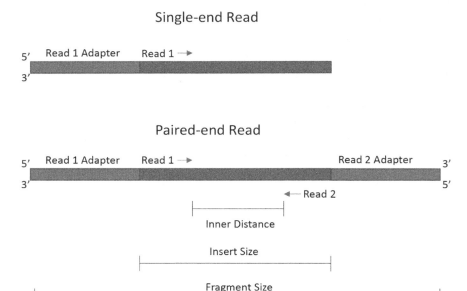

FIGURE 1.4 The general structure of single-end and paired-end reads is presented. For paired-end reads, the inner distance denotes the unknown length between the two reads, the insert size denotes the combined length of the inner distance and two paired reads, and the fragment length denotes the combined length of the entire structure including the inner distance, the two paired reads, and the two adapter sequences.

Introduction to RNA Sequencing and Quality Control 9

map to when using paired-end reads than through the use of single-end reads. In most instances, the generated fragment length is much larger than the length of the reads that are generated. The length of the reads that are generated are of known length, but the size of the fragment is not known in the case of single-end reads as only the sequence of the read is known (the entire sequence is not known from one read). In the case of paired-end reads however, we are provided increased information on the size of a fragment as we can approximate the distance between the paired reads, which aids in the accuracy of the alignment process.

1.3 RNA-seq TECHNIQUES AND PLATFORMS

1.3.1 ROCHE 454

Although the principle in which the various available sequencing platforms operate is comparable, there are advantages and disadvantages that may make one platform more preferable than another. Table 1.1 provides comparisons between the most common sequencing platforms. The previous gold standard of RNA sequencing in the early 2000s was the Roche 454 pyrosequencing platform, otherwise known as the first generation of next-generation sequencing and is a high-throughput sequencing variant of the original pyrosequencing method developed in the 1990s. In this approach, the use of streptavidin-coated magnetic beads, recombinant DNA polymerase, and detection via the enzymatic activity of firefly luciferase are combined with light intensity corresponding to the number of incorporated nucleotides. Specifically, the Roche 454 platform detects the bioluminescence signal emitted when a nucleotide is incorporated, and the remaining 2-phosphate pyrophosphate molecule is released through the enzymatic function of firefly luciferase. DNA or RNA fragments containing adapter sequences are fixed within small beads suspended in a water–oil emulsion and then are subjected to a standard PCR protocol in order to amplify the nucleotide sequences present. Each bead is then placed into a small microwell located on a fiber-optic chip, and a mixture of enzymes including polymerases and luciferase is introduced prior to placing the chip into the sequencing instrument.

Roche 454 has one primary inherent advantage that in some instances may make it more appropriate. Fragments generated by Roche 454 are typically twice as large

TABLE 1.1
Similarities and Differences between Common Sequencing Platforms

Platform	Sequencing Mechanism	Mean Read Length (bp)	Error Rate (%)	Primary Errors
Illumina HiSeq 2500	Reversible termination	125	0.26	Substitutions
Ion Torrent 318	Ligation	120	0.10	A-T bias
Roche 454	Pyrosequencing	400	0.80	Indels
SMRT	Sequencing by synthesis	1772	1.72	Mismatches
ONT	Sequencing by synthesis	1861	13.40	Mismatches

(400 bp vs. 200 bp), thus making the subsequent mapping to the reference genome a bit easier. If we use the puzzle analogy, it is easier in theory to assemble a completed genomic map from 200,000 reads than the 400,000 reads generated from Illumina.

1.3.2 ILLUMINA PLATFORM

However, in recent years, the Illumina platform has become the predominant sequencing protocol and the Roche 454 pyrosequencing platform has largely become obsolete. In the Illumina sequencing protocol, fluorescently labeled dNTPs, by which each of the four types of nucleotides is assigned a unique dye color, are incorporated into an extending DNA molecule. A dNTP is incorporated into the growing DNA strand one at a time and the color is recorded, at which point the fluorophore tag is washed away and a new fluorescent dNTP is incorporated.

Once the preparation of the cDNA library is complete, the DNA fragments are washed across the flow cell of the sequencing instrument and are exposed to primers that then retain and allow for the cDNA fragments to bind to the flow cell. The bound DNA fragments are then amplified in order to form a small cluster of DNAs with identical sequences and one that is strong enough to elicit a signal that can be detected by the instrument sensor. Next, additional unlabeled nucleotides and DNA polymerase are added to extend and connect the DNA strands bound to the flow cell in a process known as bridge amplification. Once these bridges of double-stranded DNA are constructed, the DNA is separated into single-stranded DNA using heat. Sequencing primers and the previously mentioned fluorescently labeled terminator dNTPs are then added, the primers of which are subsequently attached to the DNA strands currently being sequenced. The DNA polymerase then binds to the sequencing primers and incorporates a terminator one at a time, each of which is labeled with a different fluorescent color. A laser is then scanned across the flow cell, resulting in an output color signal for the first nucleotide. The terminator group is then removed from the first base and this process continues, in parallel, until the entirety of the sample has been sequenced.

Multiple studies have been conducted that have demonstrated the advantages and limitations of Illumina sequencing relative to the traditional Roche 454 method. One such study conducted by Luo et al. (2012) compared reads generated, error rates, and contig assemblies between the two sequencing methods for a freshwater microbial community. In sequencing, a contig is a contiguous segment of sequenced DNA that consists of multiple aligned fragments and is mapped to a reference genome. In essence, a contig is a puzzle piece that is mapped to a reference genome and is a piece of the entire genome picture of a sample. A 90% agreement in contig assembly was observed between the Roche 454 platform and the Illumina platform, and an 89% agreement for unmapped reads was found between the two platforms. In addition, estimated gene expression and transcript abundance was approximately equal between Roche 454 and Illumina sequencing. However, Illumina sequencing was found to be advantageous in several aspects. For instance, Illumina generated longer contig assemblies with a significantly lower degree of frameshift errors and indels than the reads generated from

Introduction to RNA Sequencing and Quality Control

the Roche 454 platform, particularly in A- and T-rich fragments. Finally, current Illumina sequencing is approximately 75% cheaper than the Roche 454 protocol for an equivalent sample size.

1.3.3 SMALL-SCALE RNA-SEQ PLATFORM

Another type of RNA-seq platform that is most appropriate for small-scale settings such as small laboratories and hospital settings is the Ion Torrent sequencing platform. The underlying methodology for this sequencing-by-synthesis platform is the detection of hydrogen ions that are released each time a strand of DNA is extended. A microwell that contains a template strand of DNA is exposed to one type of dNTP at a time (dATP, dGTP, dCTP, or dTTP) to check for complementarity between the template strand and the dNTP. If the dNTP is complementary to the template strand and is incorporated into the growing DNA chain, hydrogen ions will be released and trigger an ion-sensitive field-effect transistor ion sensor, which indicates a reaction had occurred through detection of an electrical signal. If the introduced dNTP is not complementary to the 3' of the growing DNA strand, then no reaction will occur, and a signal will not be detected by the sensor.

The Ion Torrent platform is advantageous for several reasons. First, operating costs are considerably lower for this platform and the cost-per-run is lower than comparable runs using previously mentioned RNA-seq platforms. Next, the machine is very rapid, can sequence samples in real time as DNA strands are extended, and several hundred nucleotides can be sequenced per hour.

However, this platform struggles in long repeat regions of the genome as the incorporation of multiple nucleotides in a single dNTP exposure event releases more hydrogen ions, resulting in a more intense signal that can make it more difficult to differentiate how many nucleotides were actually incorporated. In addition, the average read length generated from the Ion Torrent platform is limited to approximately 400 reads, thus making it less useful for whole-genome sequencing analysis. Thus, the Ion Torrent is typically used for the sequencing of microbial genomes and transcriptomes or for quality testing of sequencing libraries.

1.3.4 THIRD-GENERATION SEQUENCING

First-generation sequencing refers to technology that utilizes the Sanger chain termination method to sequence nucleotide sequences as well as bacterial cloning to construct template sequences on a small scale. Next-generation sequencing, sometimes referred to as second-generation sequencing, provides a high-throughput sequencing process that utilizes the previously described bridge PCR amplification protocol and either reversible terminator sequencing or sequencing by synthesis to prepare the template and sequence the strands, respectively.

Third-generation sequencing, including the single-molecule real-time (SMRT) platform developed by PacBio and the nanopore sequencing platform developed by Oxford Nanopore Technologies (ONT), differs from the previous generations of sequencing technology in that fragmentation and PCR amplification steps are not required during library preparation. Instead, each DNA or RNA molecule itself is directly sequenced without the need to first cut the nucleotide strands into smaller fragments.

The SMRT couples sequencing by synthesis and a fluorescent camera that emits a signal whenever a rNTP is incorporated into the growing nucleotide chain. In addition, the SMRT platform can be used to identify modified bases by measuring what is termed the interpulse duration. The interpulse duration refers to the length of time between the incorporation of two bases with longer times corresponding to nucleotides with more complex modifications. Thus, the length of time required to incorporate a given rNTP can be used to identify the type of modification present on the incorporated rNTP. The SMRT platform can generate reads greater than 900 nucleotides and costs approximately $2 per million bases.

In the ONT platform, a nanopore with bound polymerases is inserted within a membrane that is resistant to electrical pulses. An action potential is then generated in order to create a current that flows only within the nanopore. As a new nucleotide flows into the nanopore, a disruption in the electrical current, which is a specific response for each nucleotide, can be measured in order to identify the most recently incorporated nucleotide. The nanopore sequencing platform can be used to generate reads with lengths up to 98 kb and costs less than $1 per million bases.

1.4 RNA-seq FILE FORMAT

1.4.1 OUTPUT FILE: FASTQ FILE

Once the sequencing process is finished, the output of raw sequence reads will be generated and saved as a series of FASTQ files. Before delving into the intricacies of the analysis pipeline, it is necessary to identify the file types utilized at each stage of the pipeline as well as to understand the data contained within them. A FASTQ file contains 4 lines of data (Figure 1.5) for each sequenced read, which includes the following:

Line 1 starts with '@' and is succeeded by a sequence identifier label.
Line 2 contains the raw sequence data.
Line 3 always contains a '+' character.
Line 4 encodes the quality score for each nucleotide in the sequence in line 2.

The Phred quality score is a quantitative measurement of the error rate for incorrectly identifying a base in a sequence read. Each base is assigned a quality score from 0 to 40 with lower scores corresponding to a greater likelihood of misidentification of the base in question. This quality score is mathematically expressed (Eq. 1.2) as

$$Q_{score} = -10 \times \log 10 P \tag{1.2}$$

where P is the probability of misidentification of a base. In Table 1.2, the symbols, which are derived from the American Standard Coding for Information Interchange (ASCII) encoding system, and their corresponding quality scores are presented. Table 1.3 details the error rates at each increment of 10 on the Phred quality score scale.

Introduction to RNA Sequencing and Quality Control

```
@SRR12551330.1.1 1 length=76
TATTANACGCGGCCCTGTTCACTTCTCCTTCATGGTTGATCTTGATTTCTATTTCAANTTNCCCACTGACAGAACC
+SRR12551330.1.1 1 length=76
AAAAA#EEEEEEEEEEEEEEEAEEEEEEEEEEEEEEEEEEEEEEEEEEEEEEEEEEEEEEEEEEE#EE#EEAEEEEEEEEEEEE
@SRR12551330.2.1 2 length=76
TTTGTTTGTTGGCCCTGTTCACTTCTCCTTCATGGTTGATCTTGATTTCTATTTCAATTTTCCCACTGACAGAACC
+SRR12551330.2.1 2 length=76
AAAAAEEEEEEEEEEEEEEEEEEEEEEEEEEEEEEEEEEEEEEEEEEEEEEEEEEEEEEEEEEEEEEEEEEEEEEEEEE
@SRR12551330.3.1 3 length=76
AGTAGAGTGCATCAGCATACGTACTAGCATCCCCAAGTTTGTCCTGGAAGAGGGTACTGAGCTGCTGAGTGACATC
+SRR12551330.3.1 3 length=76
AAAAAEEEEEEEEEEEEEEEEEEEEEEEEEEEEEEEEEEEEEEEEEEEEEEEEEEEEEEEEEEEEEEEEEEEEEAEE
@SRR12551330.4.1 4 length=76
CATGAGCGATGAATTGGAAAGTACGGAGCCTGTATCACTCATCTGCTGCTGCTTCTCCTTCACTGGAGTCTTGAAC
+SRR12551330.4.1 4 length=76
AAAAAEEEEEEEEEEEEEEEEEEEEEEEEEEEEEEEEEEEEEEEEEEEEEEEEEEEEEEEEEEEEEEEEEEEEEEEEEE
@SRR12551330.5.1 5 length=76
TTGATCAACCATCAGCATACGTACTAGCATCCCCAAGTTTGTCCTGGAAGAGGGTACTGAGCTGCTGAGTGACATC
+SRR12551330.5.1 5 length=76
AAAAAEEEEEEEEEEEEEEEEEAEEAEEEAAAE<AE/EEEEE<AAE<AAEEEEEEEAEEEAA</EEEE/EE/EEEEEEA
@SRR12551330.6.1 6 length=76
ATTGCGGCACATCAGCATACGTACTAGCATCCCCAAGTTTGTCCTGGAAGAGGGTACTGAGCTGCTGAGTGACATC
+SRR12551330.6.1 6 length=76
AAAAAEEEEEEEEEEEEEEEEEEEEEEEEEEEEEEEEEEEEEEEEEEEEEEEEEEEEEEEEEEEEEEEEEEEEEEEEEE
@SRR12551330.7.1 7 length=76
ATGCAATTTGCTCCTAATTTCTCCAAGGTCCTGTATCCATTGCTGGGTGTGATGTCCTGTATCCATTGCTGGGTGT
+SRR12551330.7.1 7 length=76
AAAAAEEEEEEE//EEEEEEEEAEEEE/EEEEEEEEE/EAEEEEE<<EEEEEEEE</A/EEEE<EE///E/A/EA/
@SRR12551330.8.1 8 length=76
TGATGTGTTTGCATCTCCTTCCAGTGTGCCAAGGTCCACATTTGTCCCAGGGAGGGCTTCCAGGTAGTCGGGGAAG
+SRR12551330.8.1 8 length=76
AAAAAEEEEEEEEEEEEEEEEEEEEEEEEEEEEEEEEEEEEEEEEEEEEEEEEEEEEEEEEEEEEEEEEEEEEEEEEEE
@SRR12551330.9.1 9 length=76
TTAATGCAATCTGCTCGATACCCTGCAGGGTCATATCTGTCAGCATGTTGGACTCAATGCATCGCAGGAAGACATC
+SRR12551330.9.1 9 length=76
AAAAAEEEEEEEEEEEEEEEEEEEEEEEEEEEEEEEEEEEEEEEEEEEEEEEEEEEEEEEEEEEEEEEEEEEEEEEEEEA
```

FIGURE 1.5 A sample output contained in a FASTQ file. The first line begins with a @ and is followed by a sequence identified label. The second line contains the raw sequencing data for the sample, while line 3 is composed of a '+' character. Line 4 encodes a quality score for each nucleotide in line 2 and is based on the Phred quality score. Each symbol in line 4 corresponds to a particular Phred quality score. Further details on Phred quality score and its calculations are described below.

TABLE 1.2
Q-Scores and Corresponding ASCII SymbolsBelow

Symbol	ASCII Code	Q-Score
!	33	0
"	34	1
#	35	2

(Continued)

TABLE 1.2 (*Continued*)
Q-Scores and Corresponding ASCII SymbolsBelow

Symbol	ASCII Code	Q-Score
$	36	3
%	37	4
&	38	5
'	39	6
(	40	7
)	41	8
*	42	9
+	43	10
,	44	11
-	45	12
.	46	13
/	47	14
0	48	15
1	49	16
2	50	17
3	51	18
4	52	19
5	53	20
6	54	21
7	55	22
8	56	23
9	57	24
:	58	25
;	59	26
<	60	27
=	61	28
>	62	29
?	63	30
@	64	31
A	65	32
B	66	33
C	67	34
D	68	35
E	69	36
F	70	37
G	71	38
H	72	39
I	73	40

In a FASTQ data file, each nucleotide base will be assigned an ASCII symbol that corresponds to a quality score (0–40). In addition, the Q-score can be calculated by subtracting 33 from the ASCII code value.

Introduction to RNA Sequencing and Quality Control

TABLE 1.3

BelowCorresponding Misidentification Probability for Every Interval of 10 for the Phred Quality Score

Phred Quality Score	Misidentification Probability	Base Call Accuracy (%)
10	1:10	90
20	1:100	99
30	1:1000	99.9
40	1:10000	99.99

1.4.2 MAPPED FILE: SAM/BAM/BIGWIG FORMATS

The raw reads, in FASTQ format, should next be aligned and mapped to a reference genome. SAM files are another primary type of file commonly utilized in an RNA-seq pipeline that contains the short reads mapped to the reference genome. The SAM file has two distinct regions, the first of which is the header region and the second of which is the 11-line alignment section. The header region, for which a variety of required and optional tags exist, is delineated by an '@' at the beginning of each line. The information contained in each column is summarized in Table 1.4 and a sample output from Galaxy is presented in Figure 1.6.

Due to the immense size of genomic data files, SAM files can be compressed into a BAM for easier storage. For visualization purposes, particularly for the UCSC genome browser, SAM files can be further compressed into an indexed binary alignment file known as a bigwig file. The primary advantage of the bigwig format is that they are faster to display in genome browsers, largely in part due to only the portions needed to display the region of interest are transferred to the genome browser server as temporary files. The remaining portions of the file remain on your local web server. Figure 1.7 shows a portion of a bigwig file loaded into the Integrated Genome Browser.

1.4.3 GTF FILE

In addition to the FASTQ and SAM/BAM/BIGWIG files that contain the raw reads and the raw mapped sequencing data respectively, there are several other file types that are utilized concurrently for downstream analysis. The gene transfer format (GTF) is a file type that contains information on the position and type of genomic features including exons, transcripts, genes, and untranslated regions for our model organism. As shown in Figure 1.8, there are nine tab-delimited columns of information including data on chromosome location, sequence type, strand information (+ or −), sequence position, and reading frame in a GTF file. The nine columns are as follows:

   Column 1: The seq name contains the name of the chromosome or scaffold, with or without the 'chr' prefix.

16 RNA-seq in Drug Discovery and Development

TABLE 1.4
The 11 Columns for Each Template in a SAM/BAM File

Column #	Name of File Line	Description	Example Input
1	QNAME	Name of query template	Any string such as 'Chr-1'
2	FLAG	Alignment flags	256 = secondary alignment
3	RNAME	Reference sequence ID derived from fastq file.	Any character string such as 'chr2'
4	POS	1-based mapping position	32916253
5	MAPQ	Mapping quality ($-10 \log 10$) (probability mapping position is wrong).	Value between 0 and 1
6	CIGAR	Sequence of base lengths and operations	M = alignment match 7 = sequence match
7	RNEXT	Reference sequence name of the primary alignment of the next read in the template.	* = Information not available
8	PNEXT	1-based position of the primary alignment of the next read in the template.	0 = Information not available
9	TLEN	Observed length of the template.	0 = Information not available
10	SEQ	Segment sequence	GGTAGTAGGGTCCTAGG
11	QUAL	ASCII quality score	DDGDD&EFEEEFFGG%%

Column 2: Source of the data.

Column 3: The feature type name including gene, start codon, stop codon, or exon.

Column 4: Start position of the feature (the first feature on a chromosome starts with 1).

Column 5: Stop position of the feature.

Column 6: A floating point feature.

Column 7: Defines orientation of strand: + (positive) or − (negative).

Column 8: Frame identifies the reading frame of the sequence. The first base in a codon is denoted '0', the second base is denoted '1', and the third base is denoted '2'.

Column 9: The attribute section contains additional information about each feature.

In sequencing, GTF files are typically used to guide the alignment of raw reads to a reference genome so that we can identify areas of interest in which our reads mapped to. If we instead decide to align to a transcriptome, a GTF is not required, and we would simply count the number of copies of each transcript in our sample using a suitable tool.

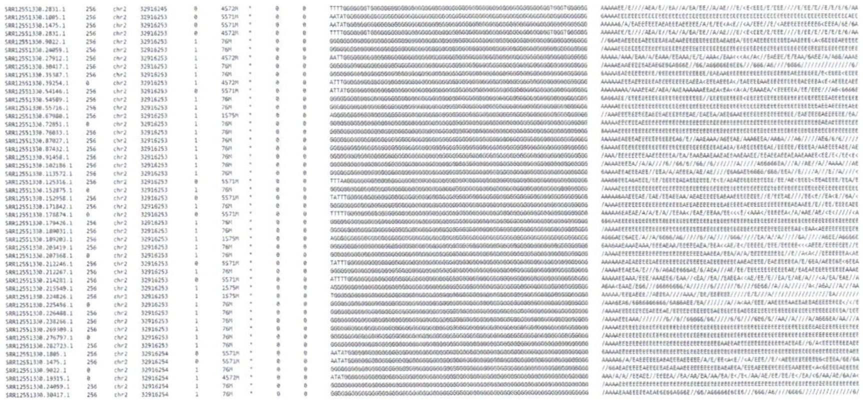

Introduction to RNA Sequencing and Quality Control 17

FIGURE 1.6 Excerpt from a SAM file generated in Galaxy.

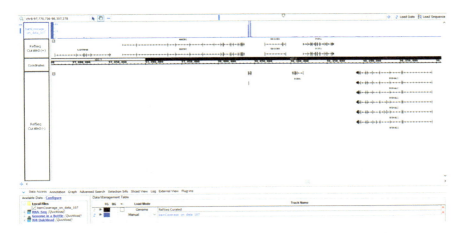

FIGURE 1.7 Sample of a bigwig file from alignment data on chromosome 8 is presented. The loaded .bam file is presented at the top of the image in blue and the reference genome is presented below the .bam file track.

1.4.4 BED FILE

Due to the significant computational power required for large genomes, GTF files are frequently converted to bed files to improve visualization in genomic browsers such as the web-based Galaxy platform, public genome browsers such as the UCSC genome browser, or a command-line tool such as BEDTOOLS. In essence, we can view a BED file, and by extension a bigBed, as the bigwig format of a GTF file that also improves browser visualization. Unlike the GTF file, the bed file is a text-based file that utilizes a zero-based coordinate system, as opposed to a nucleotide sequence, in order to describe a particular position in the genome.

18 — RNA-seq in Drug Discovery and Development

```
17  protein_coding  CDS      37866593 37866734 .         +        0        exon_number "7"; gene_biotype "protein_coding";
    gene_id "ENSG00000141736"; gene_name "ERBB2"; gene_source "ensembl_havana"; p_id "P67673"; protein_id "ENSP00000446466";
    transcript_id "ENST00000541774"; transcript_name "ERBB2-204"; transcript_source "ensembl"; tss_id "TSS58619";
17  protein_coding  CDS      37866593 37866734 .         +        0        exon_number "10"; gene_biotype "protein_coding";
gene_id "ENSG00000141736"; gene_name "ERBB2"; gene_source "ensembl_havana"; p_id "P85948"; protein_id "ENSP00000462808";
transcript_id "ENST00000578199"; transcript_name "ERBB2-003"; transcript_source "havana"; tss_id "TSS139267";
17  protein_coding  CDS      37866593 37866734 .         +        0        exon_number "7"; gene_biotype "protein_coding";
    gene_id "ENSG00000141736"; gene_name "ERBB2"; gene_source "ensembl_havana"; p_id "P68758"; protein_id "ENSP00000463714";
    transcript_id "ENST00000584450"; transcript_name "ERBB2-005"; transcript_source "havana"; tss_id "TSS69990";
17  protein_coding  CDS      37866593 37866734 .         +        0        ccds_id "CCDS32642"; exon_number "7"; gene_biotype
"protein_coding"; gene_id "ENSG00000141736"; gene_name "ERBB2"; gene_source "ensembl_havana"; p_id "P83466"; protein_id
"ENSP00000269571"; tag "CCDS"; transcript_id "ENST00000269571"; transcript_name "ERBB2-008"; transcript_source "ensembl_havana"; tss_id
"TSS171925";
17  protein_coding  CDS      37866593 37866734 .         +        0        ccds_id "CCDS45667"; exon_number "11"; gene_biotype
"protein_coding"; gene_id "ENSG00000141736"; gene_name "ERBB2"; gene_source "ensembl_havana"; p_id "P62184"; protein_id
"ENSP00000462438"; tag "CCDS"; transcript_id "ENST00000584601"; transcript_name "ERBB2-001"; transcript_source "havana"; tss_id
"TSS35358";
17  protein_coding  CDS      37866593 37866734 .         +        0        exon_number "7"; gene_biotype "protein_coding";
    gene_id "ENSG00000141736"; gene_name "ERBB2"; gene_source "ensembl_havana"; p_id "P85948"; protein_id "ENSP00000446382";
    transcript_id "ENST00000540042"; transcript_name "ERBB2-202"; transcript_source "ensembl"; tss_id "TSS128436";
17  protein_coding  CDS      37866593 37866734 .         +        0        ccds_id "CCDS45667"; exon_number "7"; gene_biotype
"protein_coding"; gene_id "ENSG00000141736"; gene_name "ERBB2"; gene_source "ensembl_havana"; p_id "P62184"; protein_id
"ENSP00000443562"; tag "CCDS"; transcript_id "ENST00000540147"; transcript_name "ERBB2-203"; transcript_source "ensembl"; tss_id
"TSS33937";
17  protein_coding  CDS      37866593 37866734 .         +        0        ccds_id "CCDS45667"; exon_number "9"; gene_biotype
"protein_coding"; gene_id "ENSG00000141736"; gene_name "ERBB2"; gene_source "ensembl_havana"; p_id "P62184"; protein_id
```

FIGURE 1.8 A sample GTF file.

In practice, a bed file works in multiple 1,000 segments in which 0 denotes the first nucleotide and 999 represents the last base in that particular segment. As a result, it becomes less time-consuming and less computationally demanding to analyze and visualize specific regions of the reference genome. The number of columns in a bed file ranges from 3 to 12, of which the first three columns are mandatory and the remaining are optional additions. Table 1.5 outlines a complete 12-column bed file with the three mandatory inputs emphasized. For columns 7–12, they may be included in the input for bedtools, but they are ignored by the program. Figure 1.9 shows a sample BED file generated using Galaxy.

TABLE 1.5
Example Inputs for Each of the 12 Columns for a Bed File

Columns	Description	Example Input	Mandatory/ Optional
Chrom	Name of chromosome or scaffold	Any string such as 'Chr-1'	Mandatory
ChromStart	Starting coordinate for the sequence of interest	Start = 10	Mandatory
ChromEnd	Ending coordinate for the sequence of interest	End = 26	Mandatory
Name	Name of the feature	Any string such as 'Exon4'	Optional

(Continued)

Introduction to RNA Sequencing and Quality Control

TABLE 1.5 (Continued)
Example Inputs for Each of the 12 Columns for a Bed File

Columns	Description	Example Input	Mandatory/Optional
Score	Score between 1 and 1,000	UCSC: Score = 800 Bedtools: Any string such as p-values, mean enrichment scores, or expression labels.	Optional
Strand	Orientation (+ or −)	'+'	Optional
ThickStart	For graphical representation, determines starting position in which annotation is more clearly defined.	12,373	Optional/ignored by bedtools
ThickEnd	For graphical representation, determines the end point in which the annotation is no longly bolded.	12,385	Optional/ignored by bedtools
ItemRgb	RGB value that determines the color of the annotations in the graphical output.	25, 500	Optional/ignored by bedtools
BlockCount	Number of blocks (e.g. exons) on the line of a bed file.	5	Optional/ignored by bedtools
BlockSizes	Comma-separated values corresponding to the size of the blocks.	360, 120, 1,222	Optional/ignored by bedtools
BlockStarts	Comma-separated values corresponding to the starting coordinates of the blocks.	4, 750, 1,403	Optional/ignored by bedtools

'Chrom', 'chromStart', and 'chromEnd' are required for every bed file, regardless of analysis software. The additional nine columns are optional inputs and are ignored if the bedtools program is used.

FIGURE 1.9 Excerpt from BED file generated from a GTF file in Galaxy.

1.5 QUALITY CONTROL OF RNA-seq DATA

1.5.1 BASIC USAGE OF THE PUBLIC SERVER GALAXY

Before we introduce the primary packages utilized in an RNA-seq pipeline, this section will introduce the basic usage of the Galaxy platform and how to access its resources. By going to usegalaxy.org and making a free account, users will have access to many different tools and reference genomes for data analysis.

The first necessary step is to upload our data, most frequently FASTQ files, into the Galaxy history so that we can access it for further downstream analysis. Figure 1.10 shows the main window for the Galaxy platform as well as the location of the upload feature. The right side of the page will contain all of the files currently available including files that were uploaded into the platform as well as any output files generated from the tools within Galaxy. The left side of the page contains a tool search bar that can be utilized to search for various tools and packages contained within the Galaxy database. Below the tool search bar is an option to upload data, either from your local or remote files. This will then allow the selected files to populate the history sidebar and will remain there until deleted by the user.

1.5.2 FastQC Program

The sequencing instrument generates raw sequencing data in the form of fastq files. As previously mentioned, there is intrinsic quality control within the sequencing instrument in the form of Phred quality scores. However, these quality scores primarily reflect issues that may arise from the instrumentation, not from possible issues of contamination or other inherent problems in the original samples. Thus, additional quality controls are typically implemented to improve

FIGURE 1.10 Main window of the Galaxy platform.

Introduction to RNA Sequencing and Quality Control

21

sample output and allow for more conclusive conclusions to be drawn from downstream analysis.

One of the most frequently used tools to supplement quality control of raw outputs includes the FastQC tool, an interactive software that can assess and identify issues in the raw output caused by both instrumentation errors and issues in the original sample and is a tool available in the Galaxy toolbox. In addition to allowing for FASTA files as input, aligned.bam and .sam files may also be used as input files into the tool.

To generate the FastQC output in Galaxy, the parameters presented in Figure 1.11 were utilized. As input for the FastQC tool, we will utilize a file containing raw reads (FASTA file format) and will leave the remaining settings at their default values for the sake of simplicity. Additional input that can be used in addition to the reads file includes a tab-delimited list of contaminants and a tab-delimited list of adaptor sequences.

Once the raw sample data has been inputted into the tool, FastQC will generate 12 different summary tabs. Figures 1.12–1.22 display an example of a high-quality output from the FastQC tool. In addition, Table 1.6 includes what will generate a 'warning message' as well as what generates a 'failure' message. The output for FastQC is an html file that links to 11 modules that describe the overall quality of the sample. In addition, each module figure will report possible warning or failure flags with a description of the issues detected. The basic *statistics report* (Figure 1.12) contains simple logistical and statistical information

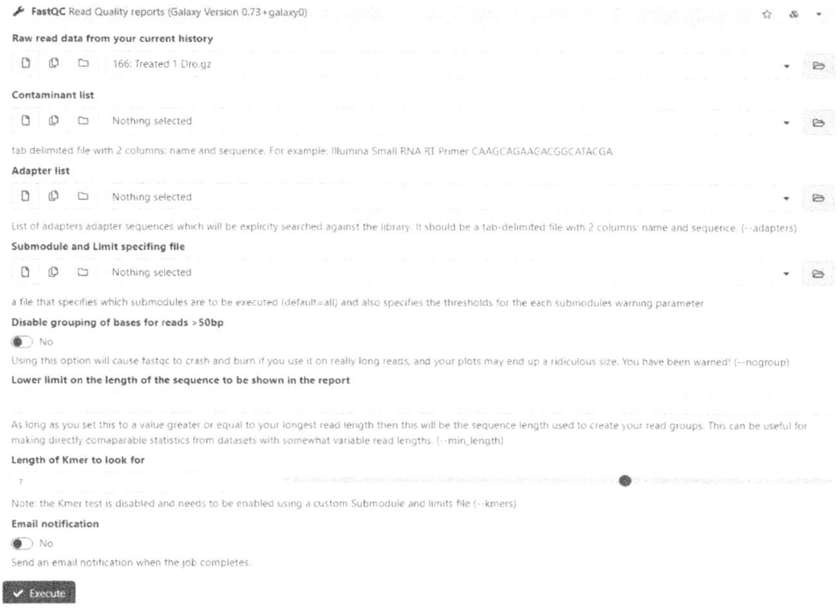

FIGURE 1.11   The parameters used to generate the FastQC report in Galaxy.

TABLE 1.6
The Purpose and Subsequent Reasons for Error Messages in Each Component of a FASTQC Quality Control Analysis

Name of Output	Characteristics	Causes of Warning Error	Causes of Failure Error
Basic statistics	Includes the name of the file, total # of sequences, sequence length, and % GC content	N/A	N/A
Per base sequence quality	Shows a BoxWhisker plot containing Phred quality scores across all bases	Lower quartile is less than 10 or the median score is less than 25	Lower quartile is less than 5 or the median score is less than 20
Per sequence quality scores	Identifies subset of sequences with poor quality	Observed mean quality lower than 27 (0.2% error rate)	Observed mean quality lower than 20 (1% error rate)
Per base sequence content	Shows the proportion of each base	Difference between A and T or G and C >10%	Difference between A and T or G and C >20%
Per sequence GC content	Measures sequence-wide GC content	Sum of the deviations from the normal distribution >15% of reads	Sum of the deviations from the normal distribution >30% of reads
Per base N content	Shows the percentage of N base calls (% of bases the instrument cannot determine)	$N > 5\%$	$N > 20\%$
Sequence length distribution	Shows distribution of fragment size	All fragments are not uniform in size	Any fragments of length$=0$
Duplicate sequences	Plots the relative numbers and duplications of each sequence	Non-unique sequences >20%	Non-unique sequences >50%
Overrepresented sequences	Lists sequences that are highly prevalent relative to other sequences	Any individual sequence is >0.1% of the total sequence population	Any individual sequence is >1% of the total sequence population
Adapter content	Determines if significant adapter contamination is present. Considered a subset of the kmer analysis	Any sequence is present in >5% of all reads	Any sequence is present in >10% of all reads
Kmer content	Shows a binomial test for even coverage of each 7-mer across the library (skipped in Galaxy)	p-value<0.01	p-value<0.00001
Per tile sequence quality	Can identify if low-quality scores are associated with a certain portion of the flow cell	Phred score more than two units below the mean in any tile	Phred score more than five units below the mean in any tile

Introduction to RNA Sequencing and Quality Control

Measure	Value
Filename	ERR4231016_1_fastq_gz.gz
File type	Conventional base calls
Encoding	Sanger / Illumina 1.9
Total Sequences	51945229
Sequences flagged as poor quality	0
Sequence length	101
%GC	50

FIGURE 1.12 The outputs for a typical FASTQC output: basic statistics.

regarding the input sample(s) including the name of the sample, total number of reads, %GC content, and the read length. In RNA-seq, GC content is important to ascertain as it may introduce bias into the experimental design. That is, bias is present if fragment counts tend to be either higher or lower in regions of the genome with high GC content.

The *per base sequence quality* module (Figure 1.13) is a box-and-whisker plot illustrating the aggregate quality score for all reads in an input FASTA file for each position. The x-axis starts out with reports for the first ten positions

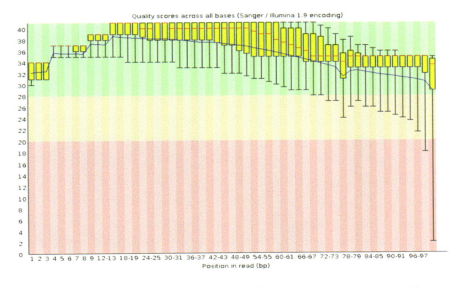

FIGURE 1.13 The outputs for a typical FASTQC output: per base sequence quality.

individually and then generates bins for the remaining base positions, the size of which is dependent on the overall read length. In general, shorter reads will be divided into bins of smaller sizes, while longer reads will be divided into larger bins. The blue line running through the graph denotes the average quality score at any given base position. For a high-quality sample, the distribution of mean read quality should reside in the upper region (green area) of the figure.

If an Illumina library retains its original sequence identifiers, another output module, termed the *per tile sequence quality* (Figure 1.14), will also be generated. Encoded by this figure is each position of the flow cell tile, ultimately allowing us to determine whether or not a loss in quality is associated with a particular area of the instrument flow cell. Overall, the figure plots the deviations in quality for each tile with cold colors (closer to blue) where the quality was at or above the average for that base in the run and warmer colors (closer to red) denoting tiles that had worse quality scores than other tiles for that particular base in the read. For high-quality samples, the plot should be consistently blue across all or most of the tiles.

The *per sequence quality score* module output (Figure 1.15) shows the statistical distribution of quality scores for all reads. Figures with the distribution heavily skewed to the right (in favor of higher quality scores) indicate a good per sequence quality sample.

The fifth output module, termed the *per base sequence content* (Figure 1.16), reports the percentage of each of the four nucleotides A, G, T, and C in each position across all reads. For high-quality samples, the %A should be approximately equal to the %T and the %G should be approximately equal to the %C.

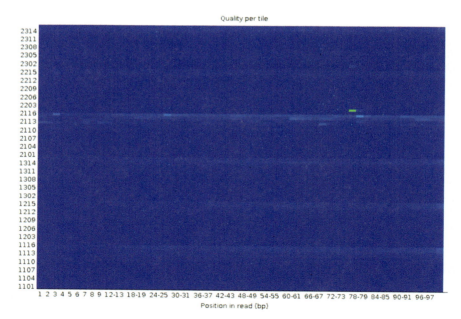

FIGURE 1.14 The outputs for a typical FASTQC output: per tile sequence quality.

Introduction to RNA Sequencing and Quality Control

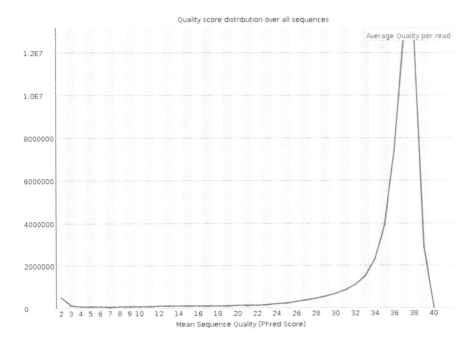

FIGURE 1.15 The outputs for a typical FASTQC output: per sequence quality scores.

It is important to note that it is very common for the first 10–15 bases to deviate from this uniform distribution.

The *per sequence GC content* output module (Figure 1.17) plots the total number of reads to the %GC content per read and displays the theoretical normal distribution vs. the distribution of our sample. We assume that our sample distribution will approximately adhere to a normal distribution; however, FASTQC is sensitive and may report a warning flag for slight deviations from the theoretical distribution despite the sample being of high quality. In many cases, this warning flag is not a cause of concern unless the theoretical and sample distributions are significantly different from one another.

The *per base N content* module (Figure 1.18) displays the percent of bases at each position that could not be called (labeled N) by the sequencing instrument. The red line in the graph should remain relatively flat around 0% and should not significantly rise above 0 across all base positions for high-quality samples.

The *sequence length distribution* module (Figure 1.19) displays the distribution of the length of the reads in the sample. For most Illumina platforms, the fragments should be of uniform size. However, many protocols and platforms generate fragments of differing sizes. As a result, many warning messages arising from this module can be ignored unless the fragment sizes differ from what is expected or what is required for your particular experiment.

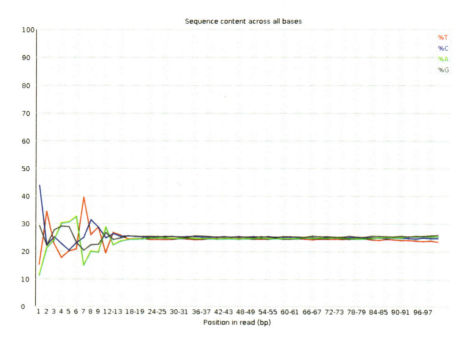

FIGURE 1.16 The outputs for a typical FASTQC output: per base sequence content.

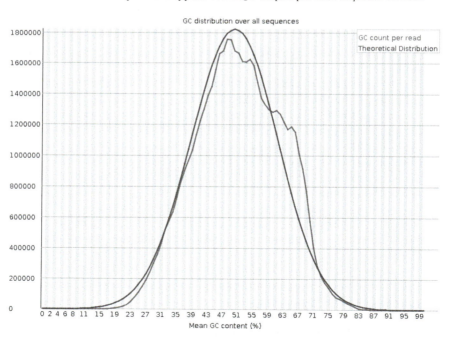

FIGURE 1.17 The outputs for a typical FASTQC output: per sequence GC content.

Introduction to RNA Sequencing and Quality Control

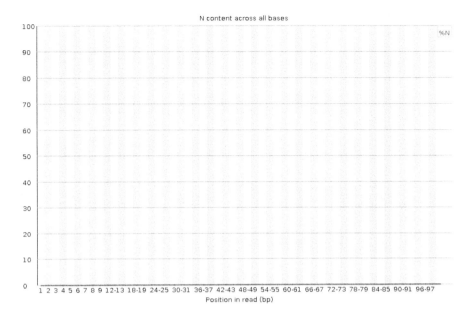

FIGURE 1.18 The outputs for a typical FASTQC output: per base N content.

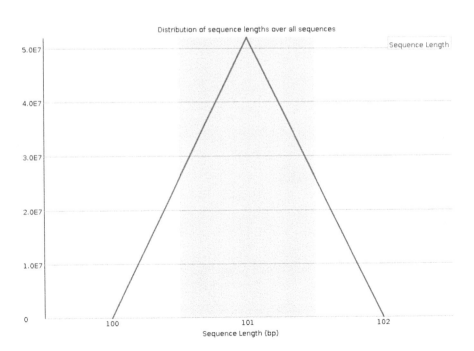

FIGURE 1.19 The outputs for a typical FASTQC output: sequence length distribution.

The *sequence duplication levels* module output (Figure 1.20) plots the percentage of reads for a given sequence in the input file to the degree of duplication of that sequence. Duplicate reads may arise either from PCR enrichment or due to the true overexpression or underexpression of a given transcript. Possible PCR enrichment misrepresents the true proportion of fragments of a given sequence in your sample and is a cause of concern; however, RNA studies are not usually concerned with this bias as it is expected that some transcripts will be present in higher or lower-than-expected abundances.

The *overrepresented* sequences module (Figure 1.21) outputs a table displaying any sequence that contributes over 0.1% of the total number of reads as well as a possible source for the overrepresentation. In many instances, adaptor sequences will be detected and included in this list.

In RNA-seq, adapters are synthetic DNA sequences that are ligated to either the 3′ end, 5′ end, or both and serve three primary purposes including

1. Contains the binding sites for the sequencing primers
2. Contains barcode and index sequences
3. Contains sites for adhering to the flow cell

The *adapter content* module (Figure 1.22) outputs a cumulative plot that displays the proportion of reads where the sequence adaptor sequence is identified

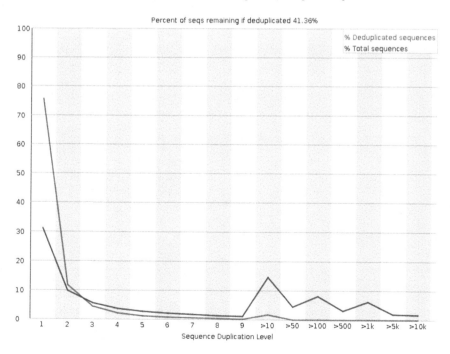

FIGURE 1.20 The outputs for a typical FASTQC output: sequence duplication levels.

Introduction to RNA Sequencing and Quality Control   29

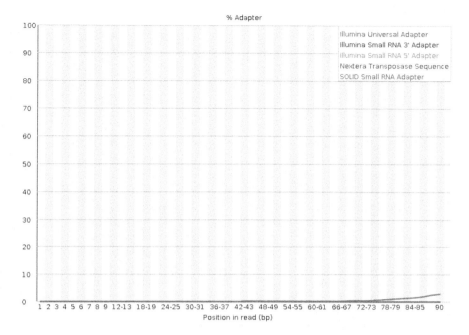

FIGURE 1.21 The outputs for a typical FASTQC output: overrepresented sequences.

Sequence	Count	Percentage	Possible Source
AGATCGGAAGAGCACACGTCTGAACTCCAGTCACCGATGTATCTCGTATG	53507	0.10300657255741429	TruSeq Adapter, Index 2 (100% over 49bp)

FIGURE 1.22 The outputs for a typical FASTQC output: adapter content.

at the indicated position. Ideally, adaptor sequences will not be detected from an Illumina library. However, when longer read lengths are used and/or the fragments are of variable sizes, it is possible that some of the library inserts will be shorter than the read length. This results in read-through to the adaptor sequence at the 3′ end of the read. Finally, because the FastQC report excludes the *kmer content* module from the output and does not perform a kmer test, we will choose to ignore this output figure.

1.5.3 TRIMMOMATIC TOOL FOR ADAPTER TRIMMING

As shown in Figure 1.22, the adapters can be identified by FASTQC. An additional quality control step that is frequently performed is adapter trimming.

30   RNA-seq in Drug Discovery and Development

In an Illumina workflow, adapter sequences are only ligated to the 3′ side of each library fragment. Thus, adapter trimming only needs to be performed on the 3′ side of each fragment when Illumina prep kits are utilized, Failure to remove these adapter sequences may result in contamination and impaired downstream analysis, particularly during the alignment phase of the pipeline. In particular, an increase in unmapped reads usually occurs, which underestimates differential expression counts. A popular tool to remove residual adapters once sequencing is complete is the Trimmomatic tool.

The Trimmomatic tool has two distinct modes that can be utilized. The 'simple mode' procedure scans each library read from 5′ to 3′ and identifies any sequences that have a high degree of similarity with the user-inputted sequences through local alignments. In addition, one can input the number of mismatches allowed so that any alignment score that exceeds the user-designated point can be removed. While this mode is very useful when the alignments are sufficiently long and the accuracy of the reads is sufficiently high, scenarios in which only small partial matches can be made reduce the ability of the tool to reliably detect the sequence of interest. The 'palindrome' mode is specifically designed to detect adaptor contamination due to 'read-through', which results in each read in a paired read having the same number of bases that are reverse complements of one another and uses global alignment scores as opposed to the local alignments assessed in 'simple mode'. 'Palindromic' mode is advantageous in that it has a longer alignment length, allowing for greater detection of adapter sequences in the presence of read errors or small adapter sequences.

1.6   ADVANTAGES OF RNA-seq OVER EXPRESSION MICROARRAYS

Although traditional microarray technology was invented in the mid-1990s, it is still frequently used in modern experiments due to its relative speed and affordability. Table 1.7 compares the major properties of both RNA-seq and expression microarrays. Microarrays are composed of thousands of small probes that are adhered to a solid substrate such as a silicon or glass slide. Each probe of the microarray contains a known sequence of synthetic, single-stranded DNA that will hybridize to a complementary sequence in the fluorescent dye labeled sample that it is exposed to. The resulting intensity of the fluorescent output signal then serves as an estimate of the expression of that particular DNA sequence or gene.

Despite the continued evolution of microarray technology, several limitations exist, which makes RNA-seq a more preferred technique. For instance, as microarrays are composed of known sequences of DNA, they are not as effective for organisms with poorly sequenced genomes or for the identification of novel transcripts and ncRNA species. In contrast, RNA sequencing does not rely on known sequences and can estimate the expression of the transcriptome in its entirety including novel isoforms or non-coding sequences. In addition, due to the repetitive nature of most genomes, cross-hybridization events may occur when utilizing

Introduction to RNA Sequencing and Quality Control

TABLE 1.7

Comparison between Expression Microarrays and RNA-seq

Property	Microarray	RNA-Seq
Requires a reference genome	Mandatory	Optional
Detection methodology	Hybridization	High-throughput sequencing
Signal output	Relative intensities	Read counts
Resolution	Length of the probe	Single nucleotide
Reproducibility	High	High
Ability to detect novel gene isoforms	No	Yes
Ability to detect SNPs	No	Yes
Range of detectable expression levels	0–300-fold-changes expression (low)	0–8,000-fold-changes expression (high)

microarrays. Cross-hybridization is defined as the binding of non-target cDNA sequences to one of the known sequences on the microarray, thus resulting in the underestimation or overestimation of expression of a particular gene. RNA sequencing thus provides a more nuanced understanding of gene expression and is more able to accurately predict the expression of a given gene.

RNA-seq is advantageous when compared to classical expression microarrays for several reasons. First, expression microarrays cannot be used to identify novel and unannotated sequences as the hybridization-based approach of the instrument is based on probes constructed from previously annotated sequences. Next, RNA-seq has a considerably higher resolution than expression microarrays with RNA-seq allowing for elucidation at the single-nucleotide level and microarrays being limited to the length of the probe used. Finally, RNA-seq can be used to identify differentially expressed genes with a significantly greater range of fold-changes in expression relative to expression microarrays. For example, expression microarrays can only identify DEGs with fold-changes between 0 and 300, while RNA-seq can be used to identify DEGs whose fold-changes are up to 8,000 or greater when compared to their baseline expression levels in untreated cells. Thus, RNA-seq provides greater flexibility and resolution than the traditional microarray-based approach.

1.7 SUMMARY

In this chapter, we introduced the concept of RNA-seq as well as the initial steps required to prepare a sample for subsequent analysis. We introduced methods of isolating RNA from DNA as well as various methods to isolate certain species of an RNA population. The major types of files used in an RNA-seq pipeline including BAM, SAM, and GTF files and their general structure were explored, and different sequencing platforms were compared. Finally, we introduced a popular tool, FastQC, to improve the quality of our raw reads files and remove low-quality reads prior to downstream analysis. In the next chapter, we will introduce various alignment tools that can utilize our raw reads and align them to a reference genome.

KEYWORDS AND PHRASES

After reading this chapter, you should be able to demonstrate familiarity with the following words and phrases:

- Understand what RNA-seq is and what purpose does it serve.
- Understand the differences between RNA-seq and microarray technology as well as advantages RNA-seq confers.
- Understand various protocols of isolating and purifying RNA from a sample including rRNA depletion, mRNA capture, and polyA selection.
- Understand what the RIN tells us.
- Be able to differentiate between RNA-seq platforms.
- Understand the difference between single-end and paired-end reads.
- Understand the major file types used in an RNA-seq pipeline and recognize their general structure.
- Understand how and why we perform quality control on our raw reads after we have received the sequencing files (fasta).
- Be able to compare and contrast high-throughput sequencing with the traditional expression microarray approach.

BIBLIOGRAPHY

Bolger, A. M., Lohse, M., & Usadel, B. (2014, Aug 1). Trimmomatic: A flexible trimmer for Illumina sequence data. *Bioinformatics, 30*(15), 2114–2120. https://doi.org/10.1093/bioinformatics/btu170

Cock, P. J., Fields, C. J., Goto, N., Heuer, M. L., & Rice, P. M. (2010, Apr). The Sanger FASTQ file format for sequences with quality scores, and the Solexa/Illumina FASTQ variants. *Nucleic Acids Res, 38*(6), 1767–1771. https://doi.org/10.1093/nar/gkp1137

Depledge, D. P., Srinivas, K. P., Sadaoka, T., Bready, D., Mori, Y., Placantonakis, D. G., Mohr, I., & Wilson, A. C. (2019, Feb 14). Direct RNA sequencing on nanopore arrays redefines the transcriptional complexity of a viral pathogen. *Nat Commun, 10*(1), 754. https://doi.org/10.1038/s41467-019-08734-9

Federico, A., Serra, A., Ha, M. K., Kohonen, P., Choi, J. S., Liampa, I., Nymark, P., Sanabria, N., Cattelani, L., Fratello, M., Kinaret, P. A. S., Jagiello, K., Puzyn, T., Melagraki, G., Gulumian, M., Afantitis, A., Sarimveis, H., Yoon, T. H., Grafstrom, R., & Greco, D. (2020, May 8). Transcriptomics in toxicogenomics: Preprocessing and differential expression analysis for high quality data. *Nanomaterials (Basel), 10*(5), 903. https://doi.org/10.3390/nano10050903

Griffith, M., Walker, J. R., Spies, N. C., Ainscough, B. J., & Griffith, O. L. (2015, Aug). Informatics for RNA sequencing: A web resource for analysis on the cloud. *PLoS Comput Biol, 11*(8), e1004393. https://doi.org/10.1371/journal.pcbi.1004393

Introduction to RNA Sequencing and Quality Control

Hoffrage, U., Krauss, S., Martignon, L., & Gigerenzer, G. (2015). Natural frequencies improve Bayesian reasoning in simple and complex inference tasks. *Front Psychol, 6*, 1473. https://doi.org/10.3389/fpsyg.2015.01473

Hong, M., Tao, S., Zhang, L., Diao, L. T., Huang, X., Huang, S., Xie, S. J., Xiao, Z. D., & Zhang, H. (2020, Dec 4). RNA sequencing: new technologies and applications in cancer research. *J Hematol Oncol, 13*(1), 166. https://doi.org/10.1186/s13045-020-01005-x

Hrdlickova, R., Toloue, M., & Tian, B. (2017, Jan). RNA-Seq methods for transcriptome analysis. *Wiley Interdiscip Rev RNA, 8*(1). https://doi.org/10.1002/wrna.1364

Kinaret, P. A. S., Serra, A., Federico, A., Kohonen, P., Nymark, P., Liampa, I., Ha, M. K., Choi, J. S., Jagiello, K., Sanabria, N., Melagraki, G., Cattelani, L., Fratello, M., Sarimveis, H., Afantitis, A., Yoon, T. H., Gulumian, M., Grafstrom, R., Puzyn, T., & Greco, D. (2020, Apr 15). Transcriptomics in toxicogenomics: Experimental design, technologies, publicly available data, and regulatory aspects. *Nanomaterials (Basel), 10*(4), 750. https://doi.org/10.3390/nano10040750

Luo, C., Tsementzi, D., Kyrpides, N., Read, T., & Konstantinidis, K. T. (2012). Direct comparisons of Illumina vs. Roche 454 sequencing technologies on the same microbial community DNA sample. *PLoS One, 7*(2), e30087. https://doi.org/10.1371/journal.pone.0030087

Minchin, S., & Lodge, J. (2019, Oct 16). Understanding biochemistry: Structure and function of nucleic acids. *Essays Biochem, 63*(4), 433–456. https://doi.org/10.1042/EBC20180038

Nyren, P., Pettersson, B., & Uhlen, M. (1993, Jan). Solid phase DNA minisequencing by an enzymatic luminometric inorganic pyrophosphate detection assay. *Anal Biochem, 208*(1), 171–175. https://doi.org/10.1006/abio.1993.1024

Quinlan, A. R., & Hall, I. M. (2010, Mar 15). BEDTools: A flexible suite of utilities for comparing genomic features. *Bioinformatics, 26*(6), 841–842. https://doi.org/10.1093/bioinformatics/btq033

Renaud, G., Stenzel, U., & Kelso, J. (2014, Oct). Adaptor trimming and merging for Illumina sequencing reads. *Nucleic Acids Res, 42*(18), e141. https://doi.org/10.1093/nar/gku699

Roberts, A., Trapnell, C., Donaghey, J., Rinn, J. L., & Pachter, L. (2011). Improving RNA-seq expression estimates by correcting for fragment bias. *Genome Biol, 12*(3), R22. https://doi.org/10.1186/gb-2011-12-3-r22

Schroeder, A., Mueller, O., Stocker, S., Salowsky, R., Leiber, M., Gassmann, M., Lightfoot, S., Menzel, W., Granzow, M., & Ragg, T. (2006, Jan 31). The RIN: An RNA integrity number for assigning integrity values to RNA measurements. *BMC Mol Biol, 7*, 3. https://doi.org/10.1186/1471-2199-7-3

Schubert, M., Lindgreen, S., & Orlando, L. (2016, Feb 12). Rapid adapter trimming, identification, and read merging. *BMC Res Notes, 9*, 88. https://doi.org/10.1186/s13104-016-1900-2

Trivedi, U. H., Cezard, T., Bridgett, S., Montazam, A., Nichols, J., Blaxter, M., & Gharbi, K. (2014). Quality control of next-generation sequencing data without a reference. *Front Genet, 5*, 111. https://doi.org/10.3389/fgene.2014.00111

Wang, Z., Gerstein, M., & Snyder, M. (2009, Jan). RNA-Seq: A revolutionary tool for transcriptomics. *Nat Rev Genet, 10*(1), 57–63. https://doi.org/10.1038/nrg2484

Weirather, J. L., de Cesare, M., Wang, Y., Piazza, P., Sebastiano, V., Wang, X. J., Buck, D., & Au, K. F. (2017). Comprehensive comparison of pacific biosciences and oxford nanopore technologies and their applications to transcriptome analysis. *F1000Res, 6*, 100. https://doi.org/10.12688/f1000research.10571.2

Xiao, T., & Zhou, W. (2020, Apr). The third generation sequencing: The advanced approach to genetic diseases. *Transl Pediatr, 9*(2), 163–173. https://doi.org/10.21037/tp.2020.03.06

Zhang, S., Wang, B., Wan, L., & Li, L. M. (2017, Jul 11). Estimating Phred scores of Illumina base calls by logistic regression and sparse modeling. *BMC Bioinform, 18*(1), 335. https://doi.org/10.1186/s12859-017-1743-4

Zhao, S., Fung-Leung, W. P., Bittner, A., Ngo, K., & Liu, X. (2014). Comparison of RNA-seq and microarray in transcriptome profiling of activated T cells. *PLoS One, 9*(1), e78644. https://doi.org/10.1371/journal.pone.0078644

Xiao, T., & Zhou, W. (2020, Apr). The third generation sequencing: The advanced approach to genetic diseases. Transl Pediatr, 9(2), 163–173. https://doi.org/10.21037/tp.2020.03.06

Zhang, S., Wang, B., Wan, L., & Li, L. M. (2017, Jul 11). Estimating Phred scores of Illumina base calls by logistic regression and sparse modeling. BMC Bioinform, 18(1), 335. https://doi.org/10.1186/s12859-017-1743-4

Zhao, S., Fung-Leung, W. P., Bittner, A., Ngo, K., & Liu, X. (2014). Comparison of RNA-seq and microarray in transcriptome profiling of activated T cells. PLoS One, 9(1), e78644. https://doi.org/10.1371/journal.pone.0078644

2 Read Alignment and Transcriptome Assembly

Robert Morris and Feng Cheng
University of South Florida

CONTENTS

2.1 Transcriptome Assembly Methodology 35
2.2 Genome-Guided Assembly 36
    2.2.1 Unspliced Aligners: Burrows–Wheeler Transform (BWT) 36
        2.2.1.1 BWT Method 37
        2.2.1.2 EXACTMATCH Method 38
    2.2.2 Unspliced Aligners: Seed Methods 40
    2.2.3 Spliced Aligners: TopHat 43
    2.2.4 Spliced Aligners: HISAT2 47
    2.2.5 Spliced Aligners: STAR 50
2.3 De Novo Assembly 53
    2.3.1 Trinity Package 55
2.4 Summary 57
Keywords and Phrases 58
Bibliography 58

2.1 TRANSCRIPTOME ASSEMBLY METHODOLOGY

Transcriptome assembly is performed by aligning the raw reads generated from the sequencing platform, usually in the form of FASTQ files, to an annotated reference genome or transcriptome of the organism in which the original cells were derived from human, mouse, *C. elegans*, etc. Many annotations can be downloaded from multiple sources including Ensembl, GENCODE, and the NCBI Reference Sequence collection (RefSeq).

Two primary methodologies exist for raw sequence alignment based on whether or not the genome is known and well annotated. The first, de novo assembly, does not require a reference genome for transcriptome assembly while the second, genome-guided assembly, utilizes a reference genome to reconstruct the transcriptome during assembly.

In addition, various tools are better at accounting for gaps in the alignment data than others. Gapped reads are reads in which splicing, insertions, deletions, or other indels are present in the sample, but not in the reference genome. Figure 2.1 illustrates a flowchart depicting the process in which a given alignment tool is

DOI: 10.1201/9781003174028-2

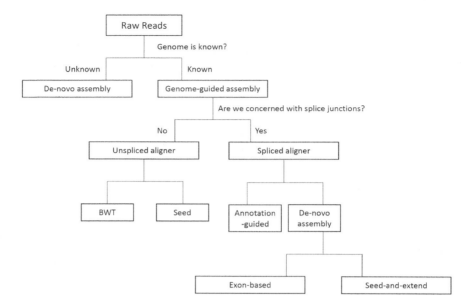

FIGURE 2.1 Two primary methodologies exist for raw sequence alignment based on whether the genome is known and well annotated.

chosen. The following sections will introduce the common algorithms behind sequence alignment for each methodology as well as commonly used tools to carry out these functions.

2.2 GENOME-GUIDED ASSEMBLY

Genome-guided assembly tools, otherwise referred to as "mapping first" assembly tools, map the raw reads to a reference genome and then assemble exons from the mapped reads. These exons are then eventually assembled as transcripts.

Genome-guided tools can be further divided into various subgroups based on whether or not gapped reads can be used, i.e., can splicing and other events such as insertions and deletions be identified. This includes short (unspliced) aligners and spliced aligners, the latter of which can be divided into annotation-guided spliced aligners and de novo spliced aligners, which can be further divided into two categories based on methodology: exon-first approaches and seed-and-extend approaches. We will first discuss unspliced aligners and their underlying workings.

2.2.1 Unspliced Aligners: Burrows–Wheeler Transform (BWT)

Unspliced aligners, sometimes referred to as short aligners, are used to align short continuous reads, which refer to reads that do not cross exon boundaries or

Read Alignment and Transcriptome Assembly 37

splice junctions. Unspliced aligners can be used to align short raw reads, usually derived from a .fastq file, to either a reference transcriptome or reference genome. Alignment to a reference transcriptome helps to bypass the limitation of identifying transcripts that span splice junctions; however, only previously annotated transcripts can be mapped to and quantified, In contrast, alignment to a reference genome allows for the identification of novel exons and transcripts. Consequently, reads that span multiple exons will more than likely not be identified and expression levels will be heavily underestimated.

2.2.1.1 BWT Method

Unspliced aligners can further be subcategorized into two categories based on their methodology. The first, known as "Burrows–Wheeler transform (BWT) methods," is utilized in such software as *Bowtie*, while the second, "seed methods," is utilized in various software including *subread*. We will first discuss BWT methods, using the software Bowtie as an example.

BWT-based short aligners utilize the BWT compression method to allow for faster and more efficient scanning of large texts, in our case a reference genome or transcriptome. The BWT transformation BWT(X) of a character string generates a matrix with the "last first (LF)" property and the rows of the matrix represent each cyclic permutation of a particular text. The rows are ordered lexicographically, which is the generalization of the alphabet to the designated order of symbols in a dataset, and the final column represents the initial character string once extracted. As indicated by the LF property, the ith position of some character X in the last column matches the ith position of some character X in the first column. The best way to illustrate this method is with an example (Figure 2.2). Using the BWT transformation of some section of the reference genome R, where R=TACGGC, we will order the nucleotides as follows: A=a (1st), C=b (2nd), G=c (3rd), and T=d (4th). BWT of a given text can be summarized in the following three steps:

1. Form all cyclic rotations of our text R=TACGGC.
2. We will append the random character \$, which is not in the character string R to BWT(X). Each cyclic rotation is then ordered alphabetically with the \$ character treated as the first letter before a. Thus, this yields seven distinct combinations of our sequence.
3. The last column in the constructed matrix is generated as our output BWT(TACGGC)=CGGCAT\$.

The BWT process is advantageous for several reasons. First, the last column of the alphabetically ordered cyclical combinations has the greatest degree of symbol clustering than any of the other columns. This stronger clustering of symbols allows for greater compression of our data and thus requires less computational memory to process. In many sequencing applications, computational requirements are significant so that the length of time needed to completely run a job is greatly reduced using compressed sequences. In addition, all possible cyclic

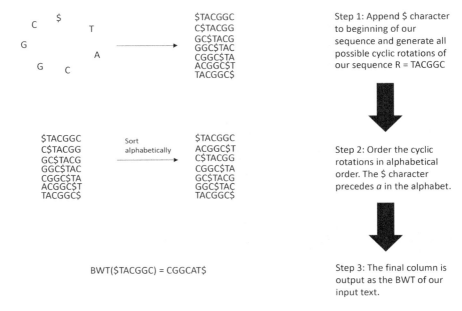

FIGURE 2.2 The generation of the BWT(X) for the sequence TACGGC

rotations can be deduced from just the BWT output of our input sequence string using the UNPERMUTATE option. As a result, this process is highly reversible and flexible.

2.2.1.2 EXACTMATCH Method

Once the BWT process is complete, the algorithm EXACTMATCH is then utilized to identify all instances of our query in our block of text (genome). EXACTMATCH calculates the range of each of the rows in the matrix with progressively longer suffixes (shortened sequences) of our query. At completion, rows that begin with our complete query, denoted S_0, correspond to exact matches of our query in the text while an empty range signifies that the query is not located in the text.

However, EXACTMATCH is not an entirely effective method for short-read aligners because of the presence of mismatches in our alignment sequences. As a result, Bowtie utilizes a modified form of the EXACTMATCH algorithm that uses a backtracking feature. The EXACTMATCH algorithm progresses as previously stated with the determination of ranges of the text (reference genome) based on progressively longer suffixes of our query. In the event that a range is empty, which occurs when a particular suffix is not present in the text, the backtrack function can add a mismatch into the alignment by identifying a previously matched query position and substituting a different base into the sequence. The EXACTMATCH algorithm then proceeds until an additional suffix is not found

Read Alignment and Transcriptome Assembly

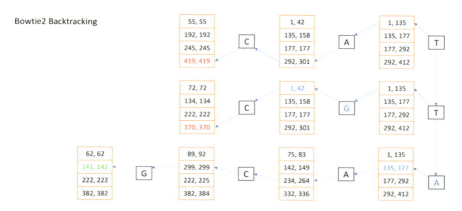

FIGURE 2.3 A sample alignment for the sequence query "TACG." Number pairs are representative of the matrix row ranges beginning with the most recent suffix observed. Red number pairs indicate points in which backtracking occurs, while blue nucleotides represent a substituted mismatch. Green number pairs represent a nonzero range that reports at least one instance of query as a reportable alignment.

within the text. When multiple candidate positions are available, the backtracking feature greedily selects a position with a low-quality value for mismatch substitution. Figure 2.3 provides an overview of the aligner algorithm utilized by Bowtie2 for the query sequence "TACG."

In some instances, the short aligner may encounter sequences that elicit excessive backtracking events. This frequently occurs when the algorithm initiates repeated backtracking to 3'-end-adjacent positions and is not completely avoidable when the number of allowed mismatches exceeds the default setting of 2. In particular, backtracking is significant when a particular read has many low-quality positions and has a poor alignment to the reference genome. To attenuate the adverse effects and increased computational workload imposed by excessive backtracking, Bowtie implements two different mechanisms. First, the user can specify an upper limit for the number of backtracking events (default setting = 125) before subsequently terminating the search query. In addition, Bowtie uses a double indexing technique that generates two indices for the reference genome, the first of which is a "forward index" with the BWT genome and a second "mirror index" that contains the reverse character sequence of the BWT genome. Both the forward index and mirror index are possible candidates for reference genome alignment.

Specifically, Bowtie allows the user to set the number of allowed alignment mismatches (default setting = two mismatches) in the high-quality end of each read, which by default is the first 28 base pairs. These 28 base pairs, termed the "seed," can be further subdivided into two distinct regions with the first 14 base pairs (by convention the 5' side) being termed the "hi-half" and the other 14 bp region being detonated the "lo-half." If we utilize the default setting of two mismatches in the seed, we get four possible reportable alignment outputs:

1. No mismatches in the 28 bp seed
2. No mismatches in the hi-half and 1–2 mismatches in the lo-half
3. 1–2 mismatches in the hi-half and no mismatches in the lo-half
4. One mismatch in both the hi-half and lo-half

In each case, the number of mismatches in the non-seed portion of each read is not of concern and a limit is not imposed by Bowtie. To identify and elucidate each of the four possible alignment outputs, Bowtie utilizes a three-phase algorithm. In phase 1, the mirror index is loaded into memory and utilizes the aligner tool to identify instances of no mismatches in the seed sequence (scenario 1) or 1–2 mismatches in the lo-half of the seed sequence (scenario 2).

Phase 2 and phase 3 of the Bowtie algorithm work in a cooperative manner to identify alignments with 1–2 mismatches in the hi-half segment of the seed sequence (scenario 3). In particular, phase 2 finds partial alignments with mismatches only allowed in the hi-half while phase 3 extends this into full alignments. Finally, phase 3 exclusively uses the alignment tool to identify alignments in which one mismatch is present in both the hi-half and lo-half segments of the seed sequence. An overview of the Bowtie phases is presented in Figure 2.4.

To generate our sample output for Bowtie2, we will use the same single-end FASTQ file that we will also as input for subsequent tools in this chapter. Although we can choose to write our aligned and unaligned reads into two separate files, we will select "no" for both parameters for the sake of simplicity. We next select whether or not we will use a reference genome that we uploaded into the history or a pre-built reference genome already available in Galaxy. For well-characterized genomes such as human or mouse genomes, the pre-built reference genomes should be sufficient. However, for less characterized or non-model organisms, a reference genome from another source or one that is built by the user may be required. For this sample output, we will utilize the pre-built human reference genome hg19 that is available in Galaxy. The remaining options will utilize the pre-selected default parameters of the Bowtie2 package. Figure 2.5 presents the parameters used by Bowtie2 to generate our sample output, while Figure 2.6 presents the sample output from Bowtie2.

2.2.2 Unspliced Aligners: Seed Methods

Seed-based methods tend to be slower than alignment tools that implore the previously mentioned BWT method; however, they also are typically more sensitive. An example of a seed-based alignment tool is subread, a general-purpose alignment tool that can be used to map both genomic DNA and RNA-seq data to a reference genome. The name of the tool subread is derived from the use of

Read Alignment and Transcriptome Assembly

FIGURE 2.4 The three steps of the Bowtie algorithm are presented above. Phase 1 identifies instances where no mismatches are present in either the hi-half or lo-half of the seed sequence (scenario 1) as well as instances where 1–2 mismatches are present in the lo-half (scenario 2). Phases 2 and 3 work in tandem to identify alignments where 1–2 mismatches are present in the hi-half of the seed sequence (scenario 3). Phase 3 utilizes the mirror index to identify alignment outputs where 1 mismatch is allowed in both the lo-half and hi-half (scenario 4).

subreads, which are shorter, equi-spaced, and overlapping pieces taken from each individual read. For the purposes of differential gene expression analysis, subread determines mapping locations by utilizing the largest mappable regions within each read. For reads that span multiple exons, subread performs local alignments to the reference genome in order to find positions with the greatest number of overlaps. As the runtime for this tool is dependent on the number of subreads utilized, the subread tool is highly scalable for increasing read lengths. The subread tool is based on the seed-and-vote algorithm, in which the subreads determine the correct alignment position for the entire read through a simple count-based vote as opposed to ranking the best matching reads. The position with the highest number of matched subreads is selected and any indels or mismatches are filled in using previously mentioned algorithms.

Mapping quality score (MQS) is calculated in a divergent manner compared to the method in which quality scores are calculated in BWT-based unspliced

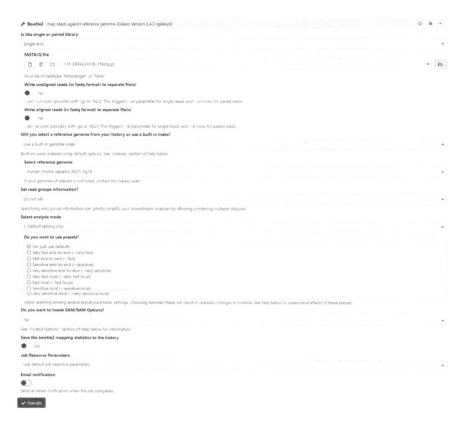

FIGURE 2.5 Parameters utilized to generate sample Bowtie2 output.

ERR4231016.	0	chr1	12022	0	86M9I6M	*	0	0	CAGCAACTG(	CCCFFFFFHHH	AS:i:-52 XS:i:-52 XN:i:0 XM:i:4 XO:i:1 XG:i:9 NM:i:13 MD:Z:84T2G0C2T0 YT:Z:UU
ERR4231016.	16	chr1	12210	1	101M	*	0	0	CTCTCACAAC	4:C@<5?A:C7	AS:i:0 XS:i:0 XN:i:0 XM:i:0 XO:i:0 XG:i:0 NM:i:0 MD:Z:101 YT:Z:UU
ERR4231016.	0	chr1	12575	1	101M	*	0	0	TGTGGGCCC/	@@@FDD1D	AS:i:0 XS:i:0 XN:i:0 XM:i:0 XO:i:0 XG:i:0 NM:i:0 MD:Z:101 YT:Z:UU
ERR4231016.	16	chr1	12658	1	101M	*	0	0	AGAGCTGCA(	3DDDDDDCC	AS:i:-6 XS:i:-6 XN:i:0 XM:i:1 XO:i:0 XG:i:0 NM:i:1 MD:Z:75G25 YT:Z:UU
ERR4231016.	0	chr1	12802	1	101M	*	0	0	GGCTTCGAT(	CCCFFFFFHHH	AS:i:-5 XS:i:-5 XN:i:0 XM:i:1 XO:i:0 XG:i:0 NM:i:1 MD:Z:86A14 YT:Z:UU
ERR4231016.	16	chr1	12906	1	101M	*	0	0	GGCAGAGTC'	DDDDDDDDD	AS:i:0 XS:i:0 XN:i:0 XM:i:0 XO:i:0 XG:i:0 NM:i:0 MD:Z:101 YT:Z:UU
ERR4231016.	16	chr1	13342	1	101M	*	0	0	CACCTTTGGT	@CDDCDDD(	AS:i:0 XS:i:0 XN:i:0 XM:i:0 XO:i:0 XG:i:0 NM:i:0 MD:Z:101 YT:Z:UU
ERR4231016.	16	chr1	13345	1	101M	*	0	0	CTTTGGTTCT	DDDDBDDDC	AS:i:0 XS:i:0 XN:i:0 XM:i:0 XO:i:0 XG:i:0 NM:i:0 MD:Z:101 YT:Z:UU
ERR4231016.	0	chr1	13414	1	101M	*	0	0	CCCCGAGAT(	CCCFFFFFHHH	AS:i:0 XS:i:0 XN:i:0 XM:i:0 XO:i:0 XG:i:0 NM:i:0 MD:Z:101 YT:Z:UU
ERR4231016.	16	chr1	13478	1	101M	*	0	0	GACAGGCAG	CCCAA?DDD(	AS:i:0 XS:i:0 XN:i:0 XM:i:0 XO:i:0 XG:i:0 NM:i:0 MD:Z:101 YT:Z:UU
ERR4231016.	16	chr1	13552	1	101M	*	0	0	GCCTGGTCT(	BDDCDDDDD	AS:i:0 XS:i:0 XN:i:0 XM:i:0 XO:i:0 XG:i:0 NM:i:0 MD:Z:101 YT:Z:UU
ERR4231016.	0	chr1	13716	1	101M	*	0	0	CCCCTCTGTC	CCCFFFFFHHH	AS:i:0 XS:i:0 XN:i:0 XM:i:0 XO:i:0 XG:i:0 NM:i:0 MD:Z:101 YT:Z:UU
ERR4231016.	0	chr1	13785	1	101M	*	0	0	CAGGAGGCT	=<+B12?DBC	AS:i:0 XS:i:0 XN:i:0 XM:i:0 XO:i:0 XG:i:0 NM:i:0 MD:Z:101 YT:Z:UU
ERR4231016.	16	chr1	14039	1	101M	*	0	0	GCTTGGTGC(	BDDDDDDDC	AS:i:0 XS:i:0 XN:i:0 XM:i:0 XO:i:0 XG:i:0 NM:i:0 MD:Z:101 YT:Z:UU
ERR4231016.	16	chr1	14079	1	101M	*	0	0	CTCTCAACCA	>9CAD?>DC(	AS:i:0 XS:i:0 XN:i:0 XM:i:0 XO:i:0 XG:i:0 NM:i:0 MD:Z:101 YT:Z:UU
ERR4231016.	0	chr1	14080	1	101M	*	0	0	TCTCAACCAC	CCCFFFFFHHH	AS:i:0 XS:i:0 XN:i:0 XM:i:0 XO:i:0 XG:i:0 NM:i:0 MD:Z:101 YT:Z:UU
ERR4231016.	0	chr1	14117	1	101M	*	0	0	CCCAACACC/	CCCFFFFFHHH	AS:i:0 XS:i:0 XN:i:0 XM:i:0 XO:i:0 XG:i:0 NM:i:0 MD:Z:101 YT:Z:UU
ERR4231016.	0	chr1	14120	1	101M	*	0	0	AACACCAGC/	CCCFFFFFHHH	AS:i:0 XS:i:0 XN:i:0 XM:i:0 XO:i:0 XG:i:0 NM:i:0 MD:Z:101 YT:Z:UU
ERR4231016.	0	chr1	14125	1	101M	*	0	0	CAGCAATTGT	CCCFFFFFHHH	AS:i:0 XS:i:0 XN:i:0 XM:i:0 XO:i:0 XG:i:0 NM:i:0 MD:Z:101 YT:Z:UU
ERR4231016.	0	chr1	14356	1	101M	*	0	0	TTTTTTTCCT(	@@@DDDD(	AS:i:-15 XS:i:-15 XN:i:0 XM:i:3 XO:i:0 XG:i:0 NM:i:3 MD:Z:3G0C0A95 YT:Z:UU
ERR4231016.	0	chr1	14358	1	101M	*	0	0	TGCATCCTGC	CCCFFFFFHHH	AS:i:0 XS:i:0 XN:i:0 XM:i:0 XO:i:0 XG:i:0 NM:i:0 MD:Z:101 YT:Z:UU
ERR4231016.	16	chr1	14373	0	4M13I4M17S	*	0	0	CAGACGTGT(	<C8BBAC<45	AS:i:-56 XS:i:-56 XN:i:0 XM:i:1 XO:i:2 XG:i:14 NM:i:15 MD:Z:0T86 YT:Z:UU
ERR4231016.	0	chr1	14410	1	101M	*	0	0	CTCAGTTCTT	CCCFFEFFHHH	AS:i:0 XS:i:0 XN:i:0 XM:i:0 XO:i:0 XG:i:0 NM:i:0 MD:Z:101 YT:Z:UU
ERR4231016.	0	chr1	14410	1	101M	*	0	0	CTCAGTTCTT	CCCFFDFFHHH	AS:i:0 XS:i:0 XN:i:0 XM:i:0 XO:i:0 XG:i:0 NM:i:0 MD:Z:101 YT:Z:UU
ERR4231016.	0	chr1	14441	1	101M	*	0	0	CTCTGGAAG(	CC:DDFFFHHH	AS:i:0 XS:i:0 XN:i:0 XM:i:0 XO:i:0 XG:i:0 NM:i:0 MD:Z:101 YT:Z:UU
ERR4231016.	16	chr1	14448	1	101M	*	0	0	AGCCTCTTAA	BBDDDDDDD	AS:i:0 XS:i:0 XN:i:0 XM:i:0 XO:i:0 XG:i:0 NM:i:0 MD:Z:101 YT:Z:UU
ERR4231016.	0	chr1	14453	1	101M	*	0	0	CTTAAGAACA	CCCFFFFFHHH	AS:i:0 XS:i:0 XN:i:0 XM:i:0 XO:i:0 XG:i:0 NM:i:0 MD:Z:101 YT:Z:UU
ERR4231016.	16	chr1	14464	1	101M	*	0	0	AGTGGCGCA	><3DBDDDC	AS:i:0 XS:i:0 XN:i:0 XM:i:0 XO:i:0 XG:i:0 NM:i:0 MD:Z:101 YT:Z:UU
ERR4231016.	0	chr1	14465	1	101M	*	0	0	GTGGCGCAG	BB@FFFFFHHH	AS:i:0 XS:i:0 XN:i:0 XM:i:0 XO:i:0 XG:i:0 NM:i:0 MD:Z:101 YT:Z:UU

FIGURE 2.6 Sample output from Bowtie.

Read Alignment and Transcriptome Assembly

aligners such as Bowtie. In subread, the following formula (Eq. 2.1) is utilized in order to assign a MQS to each mapped read:

$$MQS = 100 + \frac{100}{l} \left[\Sigma b_m \left(1 - p_i\right) - \Sigma b_{mm} \left(1 - p_i\right) \right] \qquad (2.1)$$

where l is the length of the read, p_i is the base-calling p-value for the ith base pair in the corresponding read, Σb_m is the sum of the matched base locations, and Σb_{mm} is the sum of the mismatched base locations. The base-calling p-values are derived from the raw data contained in the FASTQ files with high-quality bases subsequently having low p-values. MQS is a read-length normalized value that ranges from 0 to 200; however, if a read is best mapped to more than one location, the MQS will be divided by the number of best mapped locations.

Figure 2.7 outlines the seed-and-vote approach to mapping reads characteristic of the subread package. In this example, a read of 36 bp in length is divided into 6 bp subreads, the sequences of which are then converted into a two-bit binary combination of numbers ($A=00$, $T=01$, $C=10$, and $G=11$). These subreads are then mapped to a reference genome, and the locus of the reference genome in which the most subreads map to (highest vote count), in which no mismatches between subread and reference are allowed, is defined as the final mapping location of the read. In addition to determining the location of a read and where it maps to in the reference genome, the subread package can also identify indels present in the read.

2.2.3 Spliced Aligners: TopHat

Unspliced aligners, sometimes referred to as splice-unaware aligners, are limited by their inability to map reads that span across splice junctions. When aligning to a transcriptome, unspliced aligners are usually sufficient; however, you are ultimately limited to only identify previously identified and annotated transcripts. In organisms with less characterized transcriptomes, it is advised to avoid using a reference transcriptome as many possible existing transcripts may be unidentified and not annotated. Instead, a reference genome is preferred, which necessitates the use of a splice-aware alignment tool. Many spliced aligners will couple a splice-unaware aligner such as Bowtie with a splice-aware aligner component. Commonly utilized spliced aligners in RNA sequencing analysis include TopHat, HISAT/HISAT2, and STAR.

As previously mentioned, spice-unaware aligners are not suitable when a reference genome is used due to their inability to map reads that span splice junctions. TopHat, one of the first splice-aware aligners, uses a two-step alignment process that can also identify splice junctions. In the first phase of TopHat alignment, Bowtie is used to compare and attempt to align all reads to the reference genome. Reads that fail to align to the reference genome (spliced reads) are pooled together into a group referred to as initially unmapped (IUM) reads. For each read in the sample, Bowtie outputs one or more alignments containing less than the set

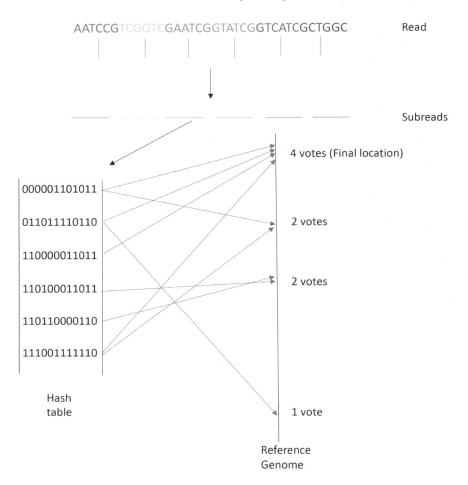

FIGURE 2.7 Seed-and-vote matching by subread package.

number of mismatches (default=2) in the 5′ end of the read. The number of alignments allowed for each read may also be adjusted (default=10). Low-complexity reads that overlap with regions of the reference genome containing small repetitive sequences are removed from the pool of IUM reads. The mapped reads are then assembled into sequence islands, which are inferred to be a putative exon, using the assemble subcommand derived from the Mapping and Assembly with Quality (MAQ) software. This yields a consensus file with called bases and their associated reference bases. Because most reads at the ends of exons will span splice junctions, and because we are initially utilizing a splice-unaware aligner to align our reads, a small amount of the end of each exon may be missing. To rectify this, as well as to identify donor and acceptor splicing sites from conjoining intronic regions, a small portion of the reference genome that flanks either side of

Read Alignment and Transcriptome Assembly 45

each of the sequence islands is included (default=45 bp). Genes with low expression may yield gapped putative exons. To combat this, TopHat has a user-definable parameter that determines the longest allowed coverage gap between islands for merging the islands into a single exon (default=6 bp).

The second phase of the TopHat pipeline is the identification of splice junctions using the reads comprising the IUM pool. First, TopHat defines all possible donor and acceptor sites, in addition to their reverse complements, within the contiguous sequence islands and then identifies which pairings of splice sites between neighboring islands could form canonical GT-AG introns. These possible intronic regions are then assessed for plausibility by comparing them to the IUM reads By default, TopHat will assess all possible exons that range in size from 70 to 20,000 bp. In order to minimize the presence of false positives in the output reports, donor–acceptor pairs located exclusively within the confines of a single sequence island are excluded unless they are highly sequenced. This is determined through the use of a D statistic, which is calculated for each sequence island during the first phase of the TopHat pipeline and measures the normalized coverage depth for each individual island. The following algorithm (Eq. 2.2) is used to compute the D value for each sequence island:

$$D_{ij} = \frac{\sum_{m=i}^{j} d_m}{j - i} \times \frac{1}{\sum_{m=0}^{j} d_m} \qquad (2.2)$$

where j and i are the coordinate range on the map for each sequence island, d_m defines the depth of coverage at a particular coordinate m, and n is the length of the reference genome. Most single-island junctions reside in islands with relatively high D values. In order to eliminate splice regions that are not deeply sequenced, as well as to improve computational running speed, TopHat only scans islands for possible junctions when $D=300$. However, this default may be lowered in order to search more sequence islands for possible splicing junctions at the cost of longer running time and greater computational demands.

TopHat uses a seed-and-extend approach when searching the IUM read pool for reads that span each splice junction. Initially, the IUM pool is indexed into a simple lookup table for easy accessibility and to minimize computational demands. For each possible splice junction, TopHat scans the index table for IUM reads, searching for any reads that span splice junctions by a specified number of k bases on either side of the junction (default=5 bp). As a result, the index table is keyed by $2k$-mers in which each $2k$-mer contains all of the IUM reads containing the particular $2k$-mer. For any given IUM read, the index table contains s-$2k$+1 possible positions for the location of a splice junction within the read where s is defined by the length of the high-quality region of the 5′ end of the read (default=28 bp). Modifications to the default settings for the variables k and s ultimately affect the sensitivity of this step of the TopHat pipeline. For example, increasing s will increase sensitivity in exchange for a longer running time while increasing k will shorten running time but hinder the identification of splice junctions in genes that are not highly expressed. A $2k$-mer seed is then generated for

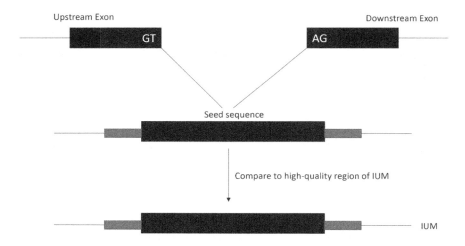

FIGURE 2.8 Seed-and-extend method utilized by TopHat.

each presumed slice junction by linking the k bases upstream of the donor site with the k bases downstream of the acceptor site. The $2k$-mer seed is then used as a query against the IUM index table to find all reads containing that particular seed and is then extended once an exact match is found.

Figure 2.8 outlines the seed-and-extend approach utilized by TopHat. Tophat utilizes many of the same parameters as Bowtie2 to generate our aligned reads. bam file. We first select whether the data input consists of single-end or paired-end reads. Next, we select our FASTQ file input as well as our reference genome, which in this instance we will utilize the built-in human genome hg19. Finally, we will leave the rest of the advanced parameters including the number of allowed mismatches, minimum intron length, and maximum intron length as their default settings and submit the job using the *execute* function. Figure 2.9 shows the parameter settings utilized to generate the sample output files.

Once the submitted TopHat software job completes, five output files are generated (Figure 2.10). These include

1. **Accepted_hits.bam**: Contains all of the read alignments.
2. **Junctions.bed**: A UCSC BED track of reported junctions. Each reported junction contains two connected BED blocks with the length of each block being determined by the maximum overhang of any read that spans the junction. The score column shows how many reads aligned to the splice junction.
3. **Insertions.bed**: A UCSC BED track of reported insertions.
4. **Deletions.bed**: A UCSC BED track of reported deletions.
5. **Align_summary.txt**: Provides overall mapping rate and the number of reads with multiple alignments.

Read Alignment and Transcriptome Assembly

FIGURE 2.9 Parameters utilized to generate sample TopHat output.

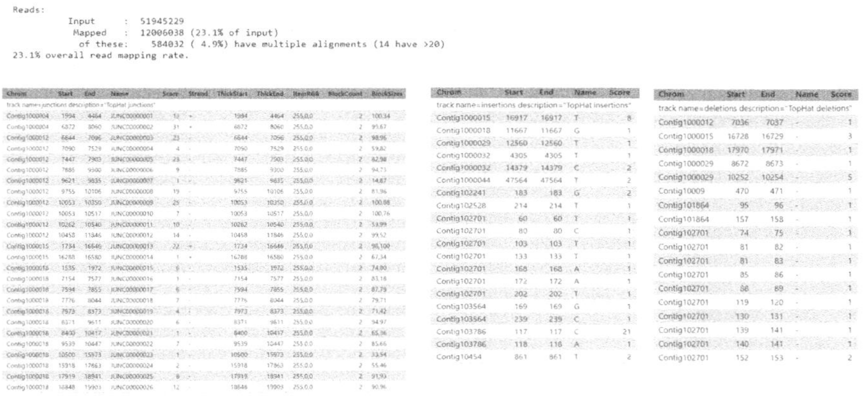

FIGURE 2.10 Sample TopHat output for alignment summary, junctions, insertions, and deletions.

2.2.4 Spliced Aligners: HISAT2

Due to the rapid advancement of the RNA sequencing field, spliced aligners have undergone multiple updates and changes to improve efficiency and accuracy. Currently, TopHat, which was first released in 2008, is in low maintenance and has been superseded most recently by hierarchical indexing for spliced alignment of transcripts two (HISAT2). The original version of HISAT utilized the BWT-based system of Bowtie2 to generate two different compressed substring indices known as FM indices. The first index is a global FM index representative of the entire reference genome and the second index being a large number of partitioned FM indices, referred to as local indices, that additively combine to represent the entire reference genome. For human genomes, approximately 48,000 FM indices are generated with each index representing 64,000 bp of the reference genome. In addition, each index overlaps its neighboring index by 1,024 bp.

While the original HISAT provided a notable improvement over TopHat in terms of efficiency and accuracy, it was, like many other aligners, limited by the reference genome itself. Human reference genomes such as GRCh38 and GRCh39 are generated from data derived from only a few individuals, thus excluding significant polymorphisms and genetic variation present in the global human population. If the source genome is very similar to the reference genome used for analysis, this does not significantly affect the accuracy of the alignment. However, if the source and reference genomes are divergent, some reads may not accurately align and overall accuracy declines. HISAT2 rectifies this by incorporating multiple single-nucleotide polymorphisms (SNPs) and polymorphisms into the reference genome by combining multiple genomic sources. First, a linear graph of the standard reference genome is created and then insertions, deletions, and SNPs are incorporated as alternative paths in the map. Similar to the de Bruijn graphs utilized in de novo assembly tools, of which will be discussed later in this chapter, each nucleotide is represented as a node and their ordering with other nucleotides is represented by edges.

Figure 2.11 shows a graphical representation of an 8 bp reference sequence (TGCAATCG) including one SNP, one deletion, and one insertion. The reference sequence (Figure 2.11a) is written out into a graph with the inclusion of one SNP (G to T), the deletion of one nucleotide (T), and the insertion of one nucleotide (G) (Figure 2.11b). The graph is then sorted lexicographically, which produces a graph with 13 nodes and 16 edges for our example (Figure 2.11c). A Z is added to represent the end of each string of nucleotides. This information is also presented in two separate tables, one of which is sorted by outgoing edges and the other is sorted by incoming edges (Figure 2.11d). In the outcoming edge table, each nucleotide is lexicographically ordered and assigned a node rank. Multiple entries in a "first" column cell correspond to a single nucleotide having multiple outgoing interconnections with subsequent nucleotides. In the incoming edge table, the preceding nucleotide is recorded for each node rank (e.g., the preceding nucleotide when node rank=7 is C). Multiple entries in a "last" column cell represent multiple nucleotides that can precede a given nucleotide in the sequence.

In HISAT2, the hierarchical indexing schematic developed for the original HISAT has been modified in order to create what is termed the Hierarchical Graph FM index. With this modification, a global index and thousands of smaller local indices are still generated; however, each local index covers 55,172 overlapping base pairs as opposed to the 48,000 base pairs seen in the original HISAT. The use of two separate graph FM indices, a reference genome and a collection of many different indel variants, allows for them to be searched and mapped to simultaneously. In addition, HISAT2 implements a novel concept of alignment in

Read Alignment and Transcriptome Assembly

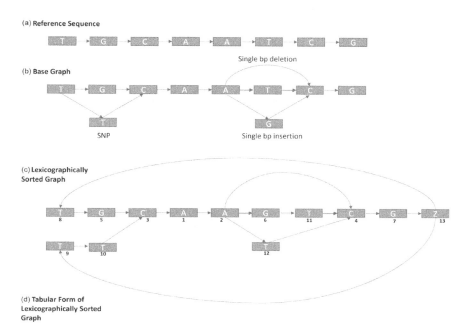

FIGURE 2.11 An example of the HISAT2 pipeline including a simple reference sequence (a), mapping of the sequence including an SNP, insertion, and deletion (b), the same map lexicographically sorted (c), and a table representation of the sorted graph.

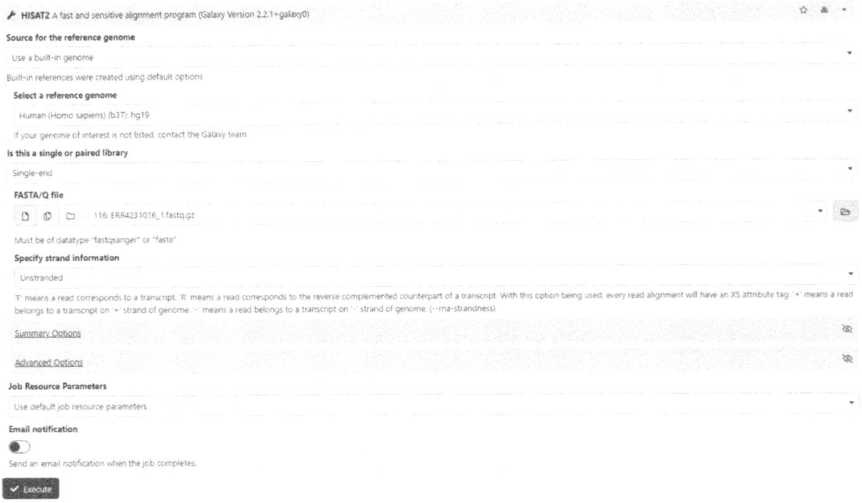

FIGURE 2.12 Parameters utilized to generate sample HISAT2 output.

order to index repeat sequences. For reads that map to many different locations (repeat sequences), many alignment tools will randomly output only a few locations due to hardware space constraints. With HISAT2, sets of identical repeat sequences in the reference genome are grouped together to form a single reference sequence that can then be directly aligned to by multiple reads. Thus, each read records one *repeat* alignment as opposed to multiple real alignments for each read in the reference genome. All of these measures previously discussed help to improve efficiency, reduce storage usage, and increase computational speed.

In order to generate our output files using HISAT2, the following parameter options will be utilized (Figure 2.12). Like the parameters for TopHat, we will select whether or not our input FASTA file will consist of single-end or paired-end reads and then select the reference genome (hg19). We will then select our input FASTQ file and select the "unstranded" option under the *specify strand information* parameter. Clicking the *execution* function will then submit the job to the queue and once completed, a .bam file of aligned reads will populate the history sidebar. A segment of the final.bam output is presented in Figure 2.13.

2.2.5 SPLICED ALIGNERS: STAR

In contrast to the previously discussed splice-aware aligners, STAR aligns non-contiguous sequences to the reference genome. STAR performs read alignment in two distinct phases, the first of which is referred to as the seed searching step and the second of which is referred to as the clustering or stitching phase. The seed searching step is fundamentally based on the step-wise search for what is termed a maximal mappable prefix (MMP). The MMP is defined as the longest substring that matches

Read Alignment and Transcriptome Assembly

FIGURE 2.13 Sample output from HISAT2.

completely to one or more substrings of the reference genome G. For a particular read sequence R and read location l, the longest substring is defined as follows:

$$R_i, R_{i+1}, \ldots, R_{i+MML-1}$$

where MML is the maximum mappable length. First, the STAR algorithm identifies the MMP by beginning with the first nucleotide of the read. For reads that contain at least one splice junction, the read cannot be mapped contiguously to the genome and thus the seed will map to a donor splice site. The algorithm will then search and repeat the aforementioned process and look for an additional MMP in the unmapped region of the read (Figure 2.14a). Although STAR is somewhat slower and more computationally intense than the previously discussed HISAT2, it is advantageous for several reasons:

1. More thorough than many other alignment tools.
2. Can more accurately identify precise locations of splice junctions in a given read.
3. Does not require a preliminary contiguous alignment pass.
4. Can identify splice junctions without a priori knowledge of their respective properties.

In addition, STAR allows for the identification and alignment of reads containing one or more mismatches. During the MMP search step of STAR, it is possible for the search to halt before reaching the end of a given read due to the presence of mismatches in the read. If this occurs, the MMPs that are generated can be used as anchors in the reference genome and subsequently extended. The MMP search occurs in both forward and reverse directions of the read sequence and the user can dictate the starting point in the read for which the search begins.

In the clustering phase of the algorithm utilized by the STAR alignment tool, alignments are constructed from the conjoining of each of the seeds that initially aligned to the reference genome (Figure 2.14b). Anchor seeds, which are

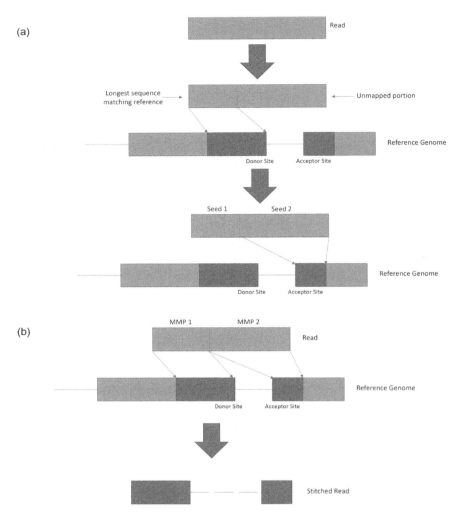

FIGURE 2.14 The seed search phase (a) and the stitching phase (b) of STAR are presented.

multimapping reads that align to the reference genome less than a user-specified number of times (default=50), are selected from the initial pool of aligned reads. Next, the reference genome is divided into equal-sized bins, also known as windows, and the remaining seeds are clustered and aligned to the reference genome. All of the non-anchor seeds that are aligned within a certain user-defined range, either upstream or downstream of the anchor seeds, are then stitched together. The user-defined range effectively dictates the longest intronic distance between the spliced alignments. If the entire read sequence is not completely covered by

Read Alignment and Transcriptome Assembly

the alignment within one genomic window, STAR will attempt to form a chimeric alignment from two or more genomic windows. The STAR algorithm can be used to locate and identify two types of chimeric alignments including

1. Chimeric alignments in which the paired mates are chimeric to one another and a chimeric junction is located in the unsequenced portion between two mates.
2. Chimeric alignments in which any or all the mates have internal chimeric alignment, thus allowing for the precise determination of the chimeric junction.

The stitching process is guided by a local alignment scoring system, of whose exact details are beyond the scope of this textbook. In essence, user-defined penalty scores are assigned to each stitched alignment for matches, mismatches, indels, and splice junctions. Stitched alignments with the highest scores are selected as the best alignment for a given read. Reads that map to multiple genomic loci will have all alignments that lie within a user-defined range reported in the output files. For sequencing technology that generates longer reads, which have higher error rates, this stitching phase is especially useful as it allows for the alignment of reads containing significant numbers of indels, mismatches, and splice junctions.

The primary output file for STAR is a .bam file containing the alignment data. In *Galaxy* and other applications that use STAR, the type of output contained in the .bam file can be adjusted. The use of the command – *outSAMtype BAM Unsorted* outputs an unsorted.bam file titled *Aligned.out.bam*, while the use of the command – *outSAMtype BAM SortedByCoordinate* will output a sorted.bam file called *Aligned.sortedByCoord.bam*. In addition, certain commands can be utilized to generate additional output files. The use of the – *outSJfilter* parameter generates a tab-delimited file called *SJ.out.tab*, which contains high confidence collapsed splice junction data. To inform STAR to detect both normal and chimeric alignments, a positive value can be assigned to the – *chimSegmentMin* option. This option dictates the minimum allowed mapped length of the two segments that will subsequently form a chimeric alignment.

2.3 DE NOVO ASSEMBLY

De novo assembly, otherwise referred to as "assembly first" methods, assembles transcripts directly from the small reads, at which point these mapped transcripts can be mapped to a reference genome if one is available. However, some limitations exist when short reads are used for this method that have made improvements in this method rather difficult including the following:

1. The determination of which reads should be joined together to form contigs.
2. Greater susceptibility to robust artifacts from the sequencing process.
3. Greater computational demands relative to genome-guided assembly.

To reduce the ambiguity associated with short-read alignment, paired-end reads can be utilized, albeit generation of paired-end reads is more costly, or the read length can be increased. In theory, assembly first methods should have greater sensitivity than genome-guided assembly. However, a combination of the presence of splicing variants, artifactual sequencing errors, and incomplete or absent reference genomes complicates the process of the read-to-reference alignment essential for de novo assembly. In addition, de novo assembly, like genome-guided sequencing, may also be plagued by complications inherent to RNA sequencing including

1. Difference in coverage between transcripts where some are lowly expressed and others are highly expressed.
2. Uneven read coverage due to sequencing biases previously discussed.
3. Reads with sequencing errors from highly expressed genes may be more abundant than correctly sequenced reads from lowly expressed genes.
4. Transcripts from adjacent loci may fuse together and erroneously be identified as a chimeric transcript.
5. The necessity of various data structures to allow for the identification of multiple transcripts at each locus (splicing variants).
6. Repetitive sequences across genes introduce ambiguity and make it more difficult to estimate gene expression.

Before we discuss the underlying algorithm utilized by most assembly first tools, we should define several relevant terms. In alignment, sequencing coverage refers to the number of read bases that align to a particular locus. In essence, a greater number of mapped bases to a single gene locus improves our confidence that a base is called correctly. The overall genomic coverage can be modified through a process known as multiplexing, a process in which multiple samples are pooled together and run simultaneously.

Many whole-genome assembly programs utilize de Bruijn graphs in order to assemble transcriptomes de novo. In a de Bruijn graph, a node is representative of a fixed length sequence of k nucleotides ("kmer") and interconnections between nodes, termed edges, are determined by perfect $k-1$ overlaps (usually 25–50 bp). When used for transcriptome assembly, each edge between nodes in a de Bruijn graph corresponds to an individual possible transcript whereas de Bruijn graphs provide a representation of all possible linear sequences that can be assembled based on the $k-1$ parameter, which is then graded on a scoring scale that utilizes the original read sequence information to remove implausible transcripts. Tools that apply the use of de Bruijn graphs in de novo transcriptome assembly must satisfy three outcomes:

1. Graphs must be constructed from raw data containing billions of base pairs.
2. Must define a proper algorithm to identify all plausible transcripts and isoforms.

Read Alignment and Transcriptome Assembly

3. Must be able to remove outside noise introduced by sequencing errors, which would generate very large de Bruijn graphs with false nodes and edges between them.

2.3.1 TRINITY PACKAGE

One of the most efficient tools for de novo transcriptome assembly is *Trinity*, a software package that contains three distinct modules: *Inchworm*, *Chrysalis*, and *Butterfly*. Figure 2.15 summarizes the procedure and output generated from each package. The first package, Inchworm, assembles the raw reads from the .fastq file into linear transcript contigs utilizing a greedy k-mer approach. Only the best representative of a set of possible alternative transcripts that share k-mers as a result of alternative splicing or allelic variation is selected. This efficient reconstruction of transcripts from linear Inchworm contigs occurs across six distinct steps.

In the first step, a *k*-mer dictionary is constructed from the raw sequence reads with the default setting $k=25$ for *k*-mer length. The second step involves the removal of *k*-mers from the Inchworm dictionary that are likely to contain sequencing errors, which is followed by the seeding of a contig assembly with the most frequently occurring *k*-mer. Next, the seed is extended in both directions through the identification of the most frequently occurring *k*-mer with a $k-1$ overlap. Once a *k*-mer is selected and the extension occurs, that *k*-mer is subsequently removed from the Inchworm dictionary. In the fifth step, the sequence is extended in either direction until it no longer overlaps and the linear contig

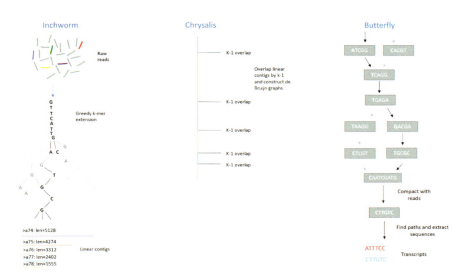

FIGURE 2.15 Outline of the three Trinity components used in de novo transcript assembly.

is then recorded. Finally, steps 3–5 are repeated until the k-mer dictionary is completely depleted. While Inchworm itself does not show the entire transcriptome, as only one alternatively spliced variant is reported in full at a particular locus, the assembled contigs contain the necessary components for Chrysalis and Butterfly to generate graphs illustrating all possible sequences.

The next component of the Trinity package, Chrysalis, groups minimally overlapping contigs from Inchworm into component sets with each component generating a de Bruijn graph. Each component is composed of all Inchworm contigs that define alternative splice variants or paralogous genes. The grouping of contigs from Inchworm and the generation of de Bruijn graphs occurs across the following three steps:

1. Grouping of contigs generated from Inchworm that share a perfect $k-1$ overlap and have a small read junction between the contigs with $(k-1)/2$ matched bases on either side of the $k-1$ junction.
2. Constructing de Bruijn graphs for each component with $k-1$ representing nodes and k defining the edges between the nodes on the graph. Each de Bruijn graph is weighted on a scale graded by the number of supporting $(k-1)$-mers from the original sequencing data.
3. Assigning a read that shares the largest number of k-mers with each component.

Now that the de Bruijn graphs are generated and we have all of the possible smaller contig sequence combinations, Butterfly is now able to construct full-length linear transcripts that utilize the original paired reads to assess plausibility. In what is referred to as the *graph simplification step*, Butterfly conjoins adjacent nodes that are on the same linear path to form nodes that are representative of longer sequences while simultaneously removing edges that are more likely to correspond to sequencing errors. The second part, *plausible path scoring*, identifies the interconnected paths that are plausible based on the original paired read data. Recall that the sequence fragments are usually several hundred base pairs while the k-mers are considerably smaller (default is $k=25$). Thus, these longer paired reads can clarify sequencing ambiguities and reduce the number of possible transcripts from the de Bruijn graphs into a smaller pool of actual transcripts. Once we have selected the Trinity tool in the Galaxy browser, we will utilize most of the default parameters of the tool. For this example, we will not be pooling sequence datasets so we will set this parameter to "no." Next, we select whether or not our input files will be single-end or paired-end reads, the former of which will be used for our sample output. We will then select the FASTQ file that we will use as our input and select "yes" for the in silico normalization of reads option. Finally, we will click the *execute* function to submit and run the job, at which upon completion will populate the history sidebar with the output files. Figure 2.16 shows a screenshot of the selected parameters used for generation of our sample output, while Figure 2.17 presents the sample output from the completed Trinity job.

Read Alignment and Transcriptome Assembly   57

FIGURE 2.16   Trinity parameters utilized to generate sample output.

FIGURE 2.17   The sample output from the completed Trinity job.

2.4   SUMMARY

In this chapter, we discussed the types of alignment tools used to align raw reads (.fastq file) to a reference transcriptome or genome. The first group, genome-guided assembly or the mapping first aligners, aligns the reads to a reference genome first to identify exons and then assembles transcripts from those exons. Mapping first aligners can be further subdivided into several distinct groups including BWT-based unspliced aligners such as bowtie2, seed-based unspliced aligners such as subread, and spliced aligners such as TopHat, HISAT2, and STAR. While de novo and unspliced aligners are sufficient when aligning to a transcriptome, spliced aligners are preferred when aligning to a reference genome due to their ability to align reads that span splice junctions. The other major group of aligners, de novo assembly alignment tools, does not require a reference genome for transcriptome assembly. De novo alignment tools such as Trinity construct transcripts directly from the reads utilizing de Bruijn graphs, at which point these transcripts can be aligned to a reference genome if one is available. In the next chapter, we will discuss commonly used packages for various downstream analyses that utilize alignment files (.bam).

KEYWORDS AND PHRASES

After reading this chapter, you should be able to demonstrate familiarity with the following words and phrases:

- Understand and be able to use the basic features of the public Galaxy server.
- Be able to describe and differentiate between BWT-based and seed-based splice-unaware methods.
- Be able to describe how the short-alignment tool Bowtie works.
- Be able to differentiate between splice-aware and splice-unaware.
- Be able to explain how TopHat works.
- Be able to explain how HISAT2 works.
- Be able to explain the underlying principles of the STAR aligner.
- Understand the concept of de novo assembly and be able to describe the underlying principles of the Trinity tool. This includes how de Bruijn graphs are constructed.
- Be able to compare and contrast de novo assembly with genome-guided assembly methods.
- Be able to use the aforementioned tools and reproduce outputs using public datasets.

BIBLIOGRAPHY

Benjamin, A. M., Nichols, M., Burke, T. W., Ginsburg, G. S., & Lucas, J. E. (2014, Jul 7). Comparing reference-based RNA-Seq mapping methods for non-human primate data. *BMC Genomics, 15*, 570. https://doi.org/10.1186/1471-2164-15-570

Dobin, A., Davis, C. A., Schlesinger, F., Drenkow, J., Zaleski, C., Jha, S., Batut, P., Chaisson, M., & Gingeras, T. R. (2013, Jan 1). STAR: Ultrafast universal RNA-seq aligner. *Bioinformatics, 29*(1), 15–21. https://doi.org/10.1093/bioinformatics/bts635

Dobin, A., & Gingeras, T. R. (2015, Sep 3). Mapping RNA-seq reads with STAR. *Curr Protoc Bioinformatics, 51*, 11–19. https://doi.org/10.1002/0471250953.bi1114s51

Dozmorov, M. G., Adrianto, I., Giles, C. B., Glass, E., Glenn, S. B., Montgomery, C., Sivils, K. L., Olson, L. E., Iwayama, T., Freeman, W. M., Lessard, C. J., & Wren, J. D. (2015). Detrimental effects of duplicate reads and low complexity regions on RNA- and ChIP-seq data. *BMC Bioinformatics, 16*(13), S10. https://doi.org/10.1186/1471-2105-16-S13-S10

Read Alignment and Transcriptome Assembly

Grabherr, M. G., Haas, B. J., Yassour, M., Levin, J. Z., Thompson, D. A., Amit, I., Adiconis, X., Fan, L., Raychowdhury, R., Zeng, Q., Chen, Z., Mauceli, E., Hacohen, N., Gnirke, A., Rhind, N., di Palma, F., Birren, B. W., Nusbaum, C., Lindblad-Toh, K., Friedman, N., & Regev, A. (2011, May 15). Full-length transcriptome assembly from RNA-Seq data without a reference genome. *Nat Biotechnol, 29*(7), 644–652. https://doi.org/10.1038/nbt.1883

Kim, D., Langmead, B., & Salzberg, S. L. (2015, Apr). HISAT: A fast spliced aligner with low memory requirements. *Nat Methods, 12*(4), 357–360. https://doi.org/10.1038/nmeth.3317

Langmead, B., & Salzberg, S. L. (2012, Mar 4). Fast gapped-read alignment with Bowtie 2. *Nat Methods, 9*(4), 357–359. https://doi.org/10.1038/nmeth.1923

Langmead, B., Trapnell, C., Pop, M., & Salzberg, S. L. (2009). Ultrafast and memory-efficient alignment of short DNA sequences to the human genome. *Genome Biol, 10*(3), R25. https://doi.org/10.1186/gb-2009-10-3-r25

Li, H., Ruan, J., & Durbin, R. (2008, Nov). Mapping short DNA sequencing reads and calling variants using mapping quality scores. *Genome Res, 18*(11), 1851–1858. https://doi.org/10.1101/gr.078212.108

Liao, Y., Smyth, G. K., & Shi, W. (2013, May 1). The subread aligner: Fast, accurate and scalable read mapping by seed-and-vote. *Nucleic Acids Res, 41*(10), e108. https://doi.org/10.1093/nar/gkt214

Liao, Y., Smyth, G. K., & Shi, W. (2019, May 7). The R package Rsubread is easier, faster, cheaper and better for alignment and quantification of RNA sequencing reads. *Nucleic Acids Res, 47*(8), e47. https://doi.org/10.1093/nar/gkz114

Trapnell, C., Pachter, L., & Salzberg, S. L. (2009, May 1). TopHat: Discovering splice junctions with RNA-Seq. *Bioinformatics, 25*(9), 1105–1111. https://doi.org/10.1093/bioinformatics/btp120

3 Normalization and Downstream Analyses

Robert Morris and Feng Cheng
University of South Florida

CONTENTS

3.1	Quantification of Transcript Abundance	61
3.2	Raw Counts Extraction	64
	3.2.1 Rsubread and *Featurecounts*	64
3.3	Normalization Methods	67
3.4	Differential Gene Expression Analysis	68
	3.4.1 *DESeq2*	68
	3.4.2 EdgeR	75
	3.4.3 Ballgown	79
3.5	Visualization of Differential Expression	81
	3.5.1 Integrative Genomics Viewer	81
	3.5.2 UCSC Genome Browser	84
	3.5.3 Heatmaps	86
	3.5.4 Volcano Plots	90
3.6	Pathway Analysis Using the NIH DAVID Web Server	91
3.7	Summary	95
Keywords and Phrases		96
Bibliography		97

3.1 QUANTIFICATION OF TRANSCRIPT ABUNDANCE

Now that we have our aligned reads, we can quantify our transcripts to estimate the gene expression which is performed by assembling and reconstructing a transcriptome from the aligned reads. One of the most commonly used assembling tools is *Stringtie*, which assembles RNA-seq alignments to form possible transcripts. The Stringtie algorithm combines a genome-guided approach as well as properties of de novo genome assembly to accurately form transcripts from aligned reads.

The input for Stringtie includes spliced-aligned reads generated from splice-aware aligners such as TopHat, HISAT2, and STAR. First, the Stringtie algorithm assembles what are termed 'super reads' from the short-aligned reads. These super reads help to bypass the higher error rate associated with longer reads as well as improve the accuracy of transcript construction by increasing the read length. In the second step, the unaligned and super reads are then aligned to the reference

DOI: 10.1201/9781003174028-3

genome in order to generate alternative splice graphs from overlapping reads. In the alternative splice graphs, each node is representative of an exon and each edge represents reads that bridge the gap between exons. Next, a flow network is created from the splice graph by following the edges with the highest weight where the weight is equal to the number of reads at that edge. The maximum flow in the network corresponds to one single transcript, which is subsequently removed from the graph and stored. A maximum flow is repeated for each transcript until no more transcripts can be assembled. After each individual assembly is complete, Stringtie's *merge* function is utilized to merge all transcripts derived from all samples into a single.gtf file. Thus, Stringtie can estimate transcript abundance for all isoforms of a gene as the weight of each edge equates to the number of reads localized to that particular gene locus.

Figure 3.1 illustrates the algorithm utilized by Stringtie, while Figures 3.2 and 3.3 show the Stringtie parameters and the subsequent sample output generated by

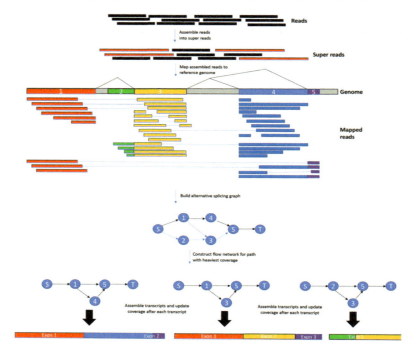

FIGURE 3.1 The process of transcriptome reconstruction employed by Stringtie. First, the individual reads are conjoined together if possible to form super reads and then both super reads and regular reads are mapped to a reference genome. Alternative splicing graphs are then built, and individual transcripts are derived based on the path with the highest weight (greatest number of mapped reads) being derived first. Once the first transcript is complete, the alternative splicing graph is updated and the path with the second highest weight is then used to define a transcript. This process is repeated until all possible transcripts are identified. Spaces between numeric positions in each individual flow diagram represent distinct exons. In the alternative splicing graph, black arrows represent the initial path with the highest weight.

Normalization and Downstream Analyses

FIGURE 3.2 Stringtie parameters utilized to generate sample output.

FIGURE 3.3 Sample output from Stringtie is presented.

Stringtie, respectively. To generate the output, we first select our aligned reads.bam file from our history. In this example, the original data was produced using the Illumina platform, which generates short reads that are less error prone. Thus, we select the 'no' option when asked whether or not the input file contains long reads. If instead the raw reads were initially produced using a different platform such as PacBio which produces longer reads that are more prone to error, we would select the

64 RNA-seq in Drug Discovery and Development

'yes' option for this parameter. We then select the 'unstranded' option for the *specify strand information* parameter and decide whether or not we should include a supplementary annotated reference file. The default setting does not require an annotated reference, which is typically sufficient for samples derived from well-characterized organisms. As our sample was derived from human cells, an annotated reference was not included to generate the output below.

3.2 RAW COUNTS EXTRACTION

The purpose of many RNA-seq experiments in drug research is to compare gene expression in cells before and after being exposed or treated with a compound of interest. To do this, it is paramount to correctly and accurately quantify gene expression. This is typically performed by assembling a count matrix from the aligned raw reads. In the count matrix, each row denotes an individual genomic feature such as exon or non-coding while each unique sample is expressed in its own column. of which the count matrix reports genomic features in rows and the different samples are expressed in the columns.

There are two commonly used tools for extraction of raw counts into a count matrix. The first, HTSeq, is a python-based program that is beyond the scope of this book, and the second, the Rsubread package, utilizes a function called *featurecounts* to construct a count matrix for genomic features of interest. For the purposes of this book, and because Rsubread is more conducive to performing subsequent downstream analyses in *R*, the next section will focus on the *featurecounts* function for extraction of raw counts.

3.2.1 RSUBREAD AND *FEATURECOUNTS*

Featurecounts is a read summarization function packaged within the unspliced aligner *subread* that can generate count matrices for genomic features of interest. There are two primary input files required for *featurecounts* to run including one or more binary alignment map (BAM) or sequence alignment map (SAM) containing aligned reads and a list of annotated genomic features in either gene transfer format (GTF) or general feature format (GFF). Single-end reads or paired-end reads can be utilized by the *featurecounts* function, and each feature of interest is designated as either a feature, which usually refers to exons or other interval-based features (regions) on one of the reference sequences, or a meta-feature, which may represent an entire gene with multiple exons or other biological constructs such as promoter regions. Depending on the *a priori* goals of the experiment, *featurecounts* may be used to count reads by gene or to count reads by exon. When counting reads by gene, each exon is considered a 'feature' and each gene, regardless of the number of exons, is termed a 'meta-feature' by Rsubread. At the gene level, each read is counted once for a gene if that read overlaps at least one exon that comprises that gene. In contrast, if Rsubread is used to count at an exon level, a read is counted for each exon that it overlaps.

The *featurecounts* algorithm works by comparing the mapping location of each nucleotide in a read to the genomic region in the reference spanned by each feature, ultimately taking into account any indels, gene fusions, or exon–exon junctions. A hit for a particular feature is called if there is at least 1 bp overlap between the read and the feature. For multi-overlap reads that span more than one feature or meta-feature, the user may choose to completely exclude these multi-overlap reads, which is typically advised for RNA-seq experiments, or they are counted for each feature or meta-feature they overlap. The first step of the *featurecounts* algorithm involves the formation of a hash table for the reference sequences, thus allowing for rapid matching between the reference sequences found in the SAM/BAM files and the reference sequences contained in the GTF/GFF annotations. Once the hash table is formed, the features are sorted by their starting positions, and a two-level hierarchy is subsequently created for each reference sequence. Initially, the reference sequences are divided into non-overlapping bins of size $n = 128\,\text{kb}$ and the features of interest are arranged among the bins based on their starting positions. An equal number of consecutive features are then grouped into blocks in each bin, where the number of blocks is equal to the $\sqrt{(\text{\# of features in bin})}$. This hierarchical data structure is beneficial as it allows for the query read to be matched first with the genomic bins followed by matching with feature blocks contained within overlapping bins. Finally, the algorithm will match the query read with features in any overlapping blocks within the bins.

Figure 3.4 summarizes the algorithm employed by *featurecounts*, and Figure 3.5 shows a sample output from the *featurecounts* function of the Rsubread package. Figure 3.6 shows the parameter options used to generate the sample output. We first select our input aligned reads.bam file and then select the 'unstranded' option under the specify strand information parameter. An annotated gene reference file can be selected from either the local history sidebar or a pre-built file provided by the *featurecounts* platform. For this sample output, aligned reads derived from mice cells were used instead of humans. Thus, we will select the pre-built mm10

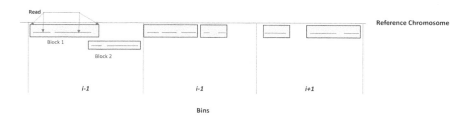

FIGURE 3.4 An outline of the *featurecounts* algorithm is presented. First, the reference genome is divided into non-overlapping bins of size $n = 128\,\text{kb}$. Next, the features of interest are sorted into the bins based on their starting position and then are grouped into blocks with other consecutive features inside the same bin. Finally, the query sequence can search for a match in a two-step process, the first of which is searching for a match within each bin and then subsequently searching for a match among the features of blocks overlapping the selected bin.

Geneid	HISAT2 on data 152: aligned reads (BAM)
Geneid	HISAT2 on data 152: aligned reads (BAM)
497097	0
100503874	0
100038431	0
19888	0
20671	0
27395	0
18777	0
100503730	0
21399	0
58175	0
108664	0
18387	0
226304	0
12421	0
620393	0
240690	0
319263	0
71096	0
59014	0
76187	0
72481	0
76982	0
17864	0
70675	0
73331	0
170755	0
620986	0
240697	0
73824	0
266793	0
100038398	0
100039596	0
69312	0
26754	0
211660	0

FIGURE 3.5 A sample *featurecounts* output is presented.

mouse genome and for the purposes of this example, will choose the output format that is compatible with DESeq2/edgeR. Finally, the *create a gene-length file option* will be set to 'no' and the remainder of the advanced parameters will be left to their default settings prior to submitting the job to the queue.

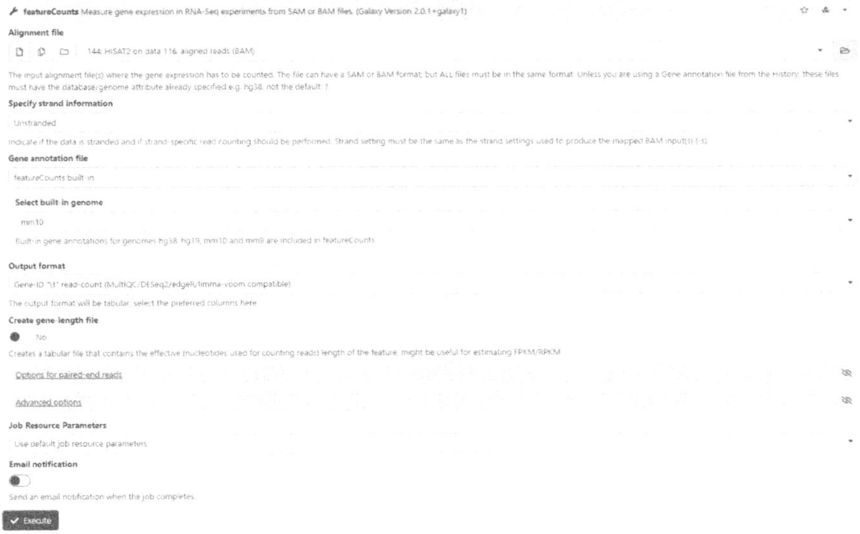

Normalization and Downstream Analyses

FIGURE 3.6 *Featurecounts* parameters utilized to generate the sample output.

3.3 NORMALIZATION METHODS

Raw counts individually do not accurately represent gene expression, especially when comparing expression across samples, due to several intervening variables including transcript length, GC content, and sequencing depth. Within-sample effects such as length and GC content alter the comparison of the gene counts of different genes within a single sample while between-sample effects such as sequencing depth prevents the comparison of the expression of the same gene across multiple conditions. Because many RNA-seq experiments are concerned with differential expression of genes across two or more conditions (samples), it is of greater importance to regulate and control for between-sample mechanistic differences. There are three primary ways to normalize raw reads including

1. **Normalization by library size**: Removes differences in sequencing depth by dividing the read count for each gene by the total number of reads in each sample.
2. **Normalization by distribution/testing**: Normalization of expression levels for non-differentially expressed genes. This assumes that technical effects are the same among both differentially expressed (DE) genes and non-differentially expressed (non-DE).
3. **Normalization by controls**: Is used when the assumptions of the other two methods are violated.

The most common way to normalize reads is normalization by library size through the generation of a measurement for each gene known as Reads per Kilobase of exon model per Million mapped reads (RPKM). The RPKM normalization method is a two-phase normalization process that takes into account both within-sample and between-sample effects. Within-sample normalization occurs due to scaling the read count for each separate transcript by the length of the transcript while between-sample normalization occurs by correcting (dividing) by the size of the library (total transcripts). If paired ends are used, we instead calculate a similar measure called Fragments per Kilobase of exon model per Million mapped reads (FPKM) where a fragment represents both pairs in a paired-end read. In addition, we can calculate the transcripts per million (TPM), which estimates the number of a particular transcript per million transcripts. The following equation can be utilized to switch between RPKM/FPKM and TPM:

$$\text{TPM} = \frac{\text{RPKM(or FPKM)} \times 10^6}{\text{sum of RPKM(FPKM)}} \tag{3.1}$$

In some cases, a few genes that are highly expressed can heavily bias the RPKM and FPKM estimates for gene expression. In this instance, we may scale the counts by the upper quantile of the counts distribution, which will be utilized by some packages used for differential gene expression analysis that will be discussed later. Regardless of the normalization method, it is important to filter and remove genes with little to no read counts. This removal of unexpressed or lowly expressed genes helps to reduce background noise, improve the power to detect differences in gene expression (reduced hypothesis testing), and improve the overall robustness of statistical significance.

3.4 DIFFERENTIAL GENE EXPRESSION ANALYSIS

Several different *R* packages exist for the assessment of DE genes across samples. However, the underlying statistical models, and thus the assumptions that are presumed to be satisfied, may differ between these packages. In addition, the choice of package for assessing differential expression will be heavily influenced by whether or not the counts generated in the previous step are pre-normalized or are raw. *R* packages such as *DESeq2* and edgeR require raw count data and thus are usually paired with *featurecounts* in a data analysis pipeline. Other packages like Ballgown require pre-normalized count data and are usually used after Stringtie. We will first discuss the underlying principles of *DESeq2* and edgeR, both of which require an input containing raw count data.

3.4.1 *DESEQ2*

The input for the *R* package *DESeq2* is a raw count matrix (k) with each row corresponding to a specific gene, denoted i, and each column corresponding to a

Normalization and Downstream Analyses

particular sample, denoted j. Each entry in the matrix, denoted k_{ij}, is equivalent to the number of reads mapped to that particular gene and each gene is fitted using a generalized linear model (GLM). Under the GLM, the read counts are assumed to follow a negative binomial distribution, a discrete probability model used for count data when the variance is larger than the mean due to overdispersion, with a mean of μ_{ij} and a dispersion of α_i. The mean is calculated by taking a quantity q_{ij}, which corresponds to the number of reads mapped to a particular gene, scaled by a normalization factor (s_{ij}). This results in the following formula (Eq. 3.2):

$$\mu_{ij} = q_{ij} \times s_{ij} \tag{3.2}$$

The scaling normalization factors are calculated using the median of the ratios of observed counts. Although in many instances the same normalization factor can be applied uniformly across all genes, it may be beneficial to calculate gene-specific normalization factors in order to capture possible biases introduced by gene length or differing GC content. For a standard case in which we are comparing gene expression between the treated and untreated samples, the GLM generated produces coefficients indicative of not only the overall expression of a gene but also the $\log_2$ fold change for the expression of every gene relative to the untreated (control) sample. Although somewhat arbitrary, a $\log_2$ fold change of >1.5 is usually used as a cutoff for significant DE genes. The general form of the GLM, which uses a logarithmic linker function, is represented by the formula (Eq. 3.3)

$$\log q_{ij} = \Sigma x_{jr} \beta_{ir} \tag{3.3}$$

where x_{jr} are matrix elements and β_{ir} are coefficients.

In *DESeq2*, within-group variability, represented by the dispersion estimate α_i, is measured by the formula (Eq. 3.4):

$$\mathrm{Var}\, K_{ij} = \mu_{ij} + \alpha_i \mu_{ij}^2 \tag{3.4}$$

Due to the small sample size of most RNA-seq experiments, dispersion estimates with high variance are assigned to each gene, which would introduce a large amount of noise into the experimental design. To rectify this, *DESeq2* utilizes a two-pronged approach that assumes genes with comparable expression levels have similar degrees of dispersion. First, a maximum likelihood dispersion estimate is calculated for each gene individually. The location parameter x_0, which is a scalar value that describes the difference in distribution of the maximum likelihood estimates of dispersion relative to a standard probability distribution of interest such as the standard normal distribution, is determined by fitting a smooth curve. This curve represents the expected dispersion values for all genes at a particular level of expression. The gene-specific dispersion estimates are then shrunk toward the expected values predicted by the curve in order to generate

70   RNA-seq in Drug Discovery and Development

final dispersion values. In essence, we are determining how far the observed (calculated) dispersion estimates differ from the expected dispersion estimates predicted by the curve. This shrinkage is done using an empirical Bayes approach, in which the strength of the shrinkage is dependent on the following two factors:

1. The estimate of how close the true dispersion values are to the predicted values of the fitted model.
2. The degrees of freedom.

As a result of this approach, there is an inverse relationship between sample size and the strength of the shrinkage. This shrinking of dispersion estimates is advantageous because it can reduce the number of false positives generated from genes with dispersion estimates significantly below the fitted curve. Conversely, for genes with gene-specific dispersion estimates at least two residual standard deviations above the expected value predicted by the fitted curve, the dispersion estimates do not undergo shrinkage and the gene-specific dispersion estimates are used as the final estimates.

In many RNA-seq experiments, comparative analysis tests the null hypothesis stating that the logarithmic fold change (LFC) between the treated samples and untreated control samples is equal to zero. In other terms, the null hypothesis states that the treatment had no statistically significant effect on the expression profile of a given gene. A negative LFC is indicative of a decline in expression of a particular gene while a positive LFC indicates overexpression of the gene in the treated samples relative to the untreated control samples. The output of this comparative analysis consists of a list of genes ranked by either adjusted p-value or LFC, in which the list can then be filtered based upon our desired LFC and significance cutoffs. Another value that may be used to identify significant genes is using the false discovery rate, a measure of the number of false rejections of the null hypothesis of no significant difference in gene expression (false positives). When ranking by adjusted p-value, which like the false discovery rate accounts for multiple hypothesis tests which would otherwise distort the true p-value, some genes with small changes in LFC may be statistically significant, but not biologically significant. As a result, *DESeq2* places more emphasis on ranking by LFC and user-inputted parameters for fold-change cutoffs to aid in the filtering of genes in which statistically significant differences in expression are not biologically relevant.

One possible obstacle faced when using LFC is the presence of significant heteroskedasticity, a measure of the variance of LFCs based only on the mean count, for lowly expressed genes. To overcome this possibility, *DESeq2* utilizes a shrinkage method that induces stronger shrinkage of genes with little information which may be due to low counts, high dispersion, or few degrees of freedom. By taking into account the amount of available information required for fold-change estimation instead of just the mean count, the amount of shrinkage of fold change estimates varies across genes. Genes with lots of available information will produce

Normalization and Downstream Analyses

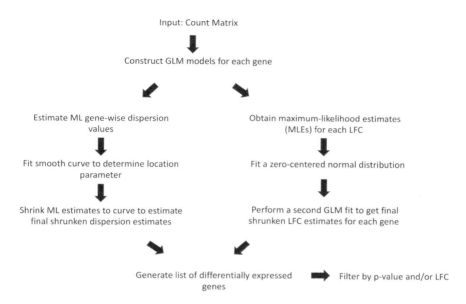

FIGURE 3.7 Outline of the *DESeq2* algorithm.

shrunken LFCs with low bias and low variance. In addition, as the number of degrees of freedom increases, the shrunken LFC estimates will converge toward the non-shrunk estimates. Figure 3.7 summarizes the algorithm employed by *DESeq2* when assessing differential gene expression between samples.

The primary output files for *DESeq2* are a gene list including *p*-values and LFCs for each gene as well as a .pdf file containing five distinct figures:

1. a principal component analysis (PCA) graph,
2. a heatmap comparing overall expression differences across samples,
3. a dispersion estimate graph illustrating the fitted curve, initial dispersion estimates, and shrunken dispersion estimates,
4. a histogram showing the distribution of *p*-values across transcripts, and
5. an MA plot showing the relationship between the log fold change and mean normalized counts across the treated and untreated samples. The name of the plot is derived from transforming the data twice, first on the log ratio scale (denoted M) and next on the mean average scale (denoted A).

PCA is a dimensionality reduction method used to take multivariate data and present the data in such a way that it is more easily interpretable. To do this, PCA involves plotting the data points on the few most important principal components that preserve as much of the observed variance as possible. A principal component can be defined as a dimensional direction that maximizes the variance of the plotted data points. Heatmaps present differential gene expression in terms of color gradients and may be utilized for gene-specific gene expression or overall differential gene expression across samples. For example, as genes undergo reduced expression, the quadrant corresponding to the specific gene and sample will increasingly turn more blue while overexpression will result in quadrants becoming a darker shade of red. Heatmaps and other methods of differential gene expression presentation, as well as how to generate them in Galaxy, will be explored in more detail later in this chapter, but for now, we will explain the specific outputs generated by *DESeq2*.

To explain these figures, we will utilize a public dataset from NCBI (GEO accession: GSE157167) and generate sample outputs from two separate datasets, each of which has two replicates. The first experimental design will assess differential gene expression in wildtype (WT) mouse intestinal cells treated with the antihelminthic drug albendazole vs. an untreated WT control while the second design will assess differential gene expression in AKP mutant mouse intestinal cells treated with albendazole relative to untreated AKP intestinal cell controls. Figure 3.8a presents a sample of the differential gene expression list generated for the WT samples ranked by adjusted *p*-value. As evident by the list, no statistically significant difference in gene expression was detected when comparing the transcriptome profiles of WT cells treated with albendazole and untreated WT cells. In contrast, multiple statistically significant genes were determined to be differentially expressed in treated AKP intestinal cells when compared to untreated controls (Figure 3.8b).

Figures 3.9 and 3.10 present the five output files for the WT and AKP differential gene expression analysis. The sample-to-sample distance heatmaps

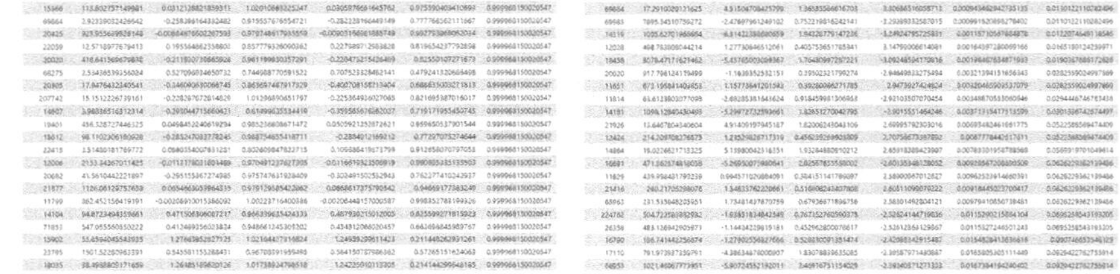

FIGURE 3.8 *DESeq2* gene list output for WT treated vs. untreated (a) and AKP treated vs. untreated (b).

Normalization and Downstream Analyses

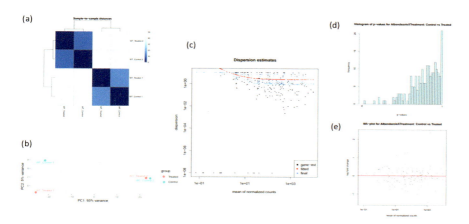

FIGURE 3.9 *DESeq2* output for WT treated vs. untreated cells including the sample-to-sample heatmap comparing gene expression (a), PCA plot (b), plot showing initial and fitted estimates of dispersion (c), histogram showing distribution of *p*-values (d), and MA plot that shows the relationship between LFC change in expression and the mean expression strength (e).

compare the overall expression profiles across samples and allow us to visualize the similarities and differences between the samples. In these graphs, quadrants with darker shades of blue are indicative of stronger similarities between the two specific samples, while lighter shades are indicative of increasing divergence of samples. For the WT heatmap (Figure 3.9a), the two untreated controls are significantly different from one another on a global scale and the two treated samples are significantly divergent from each other. In addition, the second treated sample is quite similar to the second untreated sample. Despite these global similarities and differences, it does not necessarily mean these will be observed on a gene level, such that statistically significant genes may exhibit patterns in which the controls and treated samples are more comparable to their replicates than the global pattern suggests.

For the AKP samples (Figure 3.10a), the untreated controls were very similar to one another while the two treated samples had significantly divergent global expression patterns from each other. This clustering of samples is also observed in the PCA plots for both the WT (Figure 3.9b) and AKP (Figure 3.10b) samples. In the dispersion estimate plots for the WT samples (Figure 3.9c) and AKP samples (Figure 3.10c), black dots are representative of the preliminary gene-wise estimates of dispersion, red dots comprise the fitted smooth curve, and the blue dots are representative of the final shrunken estimates of dispersion for each gene. When comparing the *p*-value histogram of the WT samples (Figure 3.9d) with the histogram generated for the AKP samples (Figure 3.10d), the distribution of the

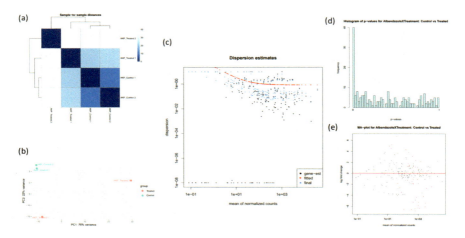

FIGURE 3.10 *DESeq2* output for AKP treated vs. untreated cells including the sample-to-sample heatmap comparing gene expression (a), PCA plot (b), plot showing initial and fitted estimates of dispersion (c), histogram showing distribution of p-values (d), and MA plot that shows the relationship between LFC change in expression and the mean expression strength (e).

p-values is skewed heavily to the right, suggesting that most of the corresponding *p*-values are not statistically significant. This is in agreement with the gene list that was also generated by *DESeq2* as no statistically significant difference in gene expression was detected. In contrast, the *p*-value distribution for the treated AKP cells is heavily skewed to the left, thus suggesting that most of the gene-specific *p*-values reached the required level of significance.

Finally, the MA plots express the comparison of the LFC (*y* axis) to the mean of normalized counts (*x* axis) across samples. Because we assume in an RNA-seq experiment that most genes are not differentially expressed, most of the genes should lie around the horizontal line as log(0) = 1. Under this assumption, genes with similar levels of abundance across samples should have comparable LFCs and should be plotted in similar positions on an MA plot. DE genes, denoted by a red dot corresponding to each gene on the MA plot, thus have statistically significant differences between their actual plotting location and the expected relationship between mean abundance and LFC. For the WT MA plot (Figure 3.9e), most of the genes are clustered around the horizontal line, and those that cluster further away from log(0) = 1 do not achieve statistical significance. In contrast, there is considerably less clustering of genes close to the horizontal line in the AKT samples and many genes reach a statistically significant difference in expression relative to their abundance (Figure 3.10e).

To generate the previous figures using the *DESeq2* platform in Galaxy, the following parameters were utilized (Figure 3.11). For the above example, two replicates were used for each treated and control groups. The *DESeq2* platform must

Normalization and Downstream Analyses

FIGURE 3.11 *DESeq2* parameters for WT treated vs. untreated (a) and AKP treated vs. untreated (b).

ultimately be run twice; once for the WT treated vs. control (Figure 3.11a) and again for the AKP treated vs. control (Figure 3.11b). Under the *how* parameter, we will first select the 'select datasets per level' option. We must then label our factor(s) and the corresponding levels for each factor. In this instance, a factor corresponds to any distinguishing characteristic or variable between our datasets. In this example, our 1 factor will be albendazole treatment status and it will be split into two levels. The first level will specify the untreated group and the corresponding input will be the individual *featurecounts* output files generated for each control sample. In order to select multiple files from the dropdown menu, hold the *ctrl* button and select the appropriate files. The second level will specify the treated group and the input will be each of the *featurecounts* output files corresponding to the albendazole-treated group. Additional factors may be incorporated depending on the constraints of the experimental design. As we are using raw count data from *featurecounts*, we will select 'yes' when asked whether or not our input files contain file headers in the first row of the files as well as select the 'count data' option for the *choice of input* parameter. Once the job is completed, two separate output files will populate the history section for each comparison.

3.4.2 EDGER

Like *DESeq2*, the input for the *R* package edgeR is a raw count matrix (k) with each row corresponding to a specific gene, which is denoted i, and each column corresponding to a particular sample, denoted j. Each entry in the matrix, denoted k_{ij}, is equivalent to the number of reads mapped to that particular gene and each gene is fitted using a generalized linear model (GLM. However, contrary to the assumption that the read counts follow a negative binomial distribution as is the case when using *DESeq2*, the read counts input for edgeR is allowed to also assume to follow an overdispersed Poisson distribution, sometimes referred to as a quasi-Poisson distribution. Although both the negative binomial distribution and overdispersed Poisson distribution are commonly used for count data and frequently provide comparable model parameter coefficients, the relationship between the mean and variance differs between the two. For an overdispersed (quasi) Poisson distribution, the variance is a linear function of the mean. In contrast, the variance is a quadratic function of the mean in negative binomial models. The primary determinant of which model class to utilize is based on how well the model will fit the observed data, a factor of which is determined by whether or not the relationship between the predictors (x variables) and response variable (y) is primarily a linear association or a more complex one with a weaker linear component. In essence, the Poisson distribution is a special case of the negative binomial distribution where the mean and variance are assumed to be equal. Thus, an overdispersed Poisson distribution is an intermediate between a typical Poisson distribution and a negative binomial distribution as a model with an overdispersed Poisson distribution contains an extra parameter that can account for dispersion.

With this in mind, edgeR fits a negative binomial distribution model with the following form (Eq. 3.5):

$$Y_{gi} \sim NB\left(M_i p_{gj}, \phi_g\right) \tag{3.5}$$

where M_i refers to the total number of reads, ϕ_g denotes the dispersion, and p_{gj} is the relative abundance of the gene g in experimental group j to which sample i belongs. The mean (Eq. 3.6) and variance (Eq. 3.7) can be calculated using the following formulas:

$$\mu_{gi} = M_i p_{gj} \tag{3.6}$$

$$var = \mu_{gi}\left(1 + \mu_g \phi_g\right) \tag{3.7}$$

An important measure for edgeR to estimate is the biological coefficient of variation (BCV), which is an estimate of the variation of the true abundance of genes between samples if all technical variation could be reduced to zero by increasing sequencing depth indefinitely. Let π_{gi} denote the proportion of all mapped reads/fragments from a gene g within the ith sample and let G represent the total number of genes so that the $\Sigma G_{g=1} \times \pi_{gi} = 1$. In addition, let $\sqrt{\phi_g}$ denote the biological coefficient of variation (BCV of π_{gi} between i replicates, which is calculated by dividing the standard deviation by the mean, N_i denote the total number of mapped reads in sample i, and y_{gi} denote the number of reads that map to a gene

Normalization and Downstream Analyses

g. We can then calculate the expected value of y_{gi} (mean) using the following formula (Eq. 3.8):

$$E(y_{gi}) = \mu_{gi} = N_i \pi_{gi} \tag{3.8}$$

We can then calculate the CV^2 value using the following formula (Eq. 3.9):

$$CV^2(y_{gi}) = \frac{1}{\mu_{gi}} + \phi_g \tag{3.9}$$

The manner of which edgeR calculates the dispersion estimates is dependent on whether or not the original experimental design separated samples by a single factor (e.g. treated vs. untreated) or multiple factors. For single-factor experiments, the quantile-adjusted conditional maximum likelihood (qCML) is utilized, which determines the likelihood conditioned on the total counts for each gene id (transcript, exon, tag, etc.) by using pseudocounts. Pseudocounts are representative of the expected normalized counts under the assumption of equal-sized libraries and are used internally by the edgeR platform to decrease the computational time required. Given a table of counts, qCML can be used to calculate both total common dispersion and individual gene-wise dispersion estimates. We can then test for differential gene expression by using the *exact test* function to calculate *p*-values by summing the sum of all of the counts with a probability smaller than the probability assumed under the null hypothesis of the observed sum of counts.

For more complex experimental designs involving multiple factors, the qCML method is insufficient in accurately estimating the dispersion parameters as it does not consider the effects and interactions between multiple factors. Thus, an approximate conditional likelihood-based method known as Cox–Reid profile-adjusted likelihood is used instead, which takes the effects of multiple factors into account. When provided with a count table and a matrix detailing the experimental design, edgeR fits log-linear models for each gene in order to calculate common dispersion among all gene ids, dispersion trends, and gene-wise dispersion estimates. Each log-linear model fits the following general formula (Eq. 3.10):

$$\log \mu_{ij} = x_i \beta_g + \log N_i \tag{3.10}$$

where x_i denotes a vector that describes the treatment conditions for a given sample *i* and β_g denotes a vector describing the regression coefficients by which the covariate effects are mediated for a particular gene *g*. Hypothesis testing can then proceed using an empirical Bayes quasi-likelihood (QL) F-test which provides a more robust error rate control than a standard likelihood ratio test for experiments with a small number of replicates.

The following edgeR sample output uses data from the same public dataset (GEO accession: GSE157167), but instead assesses differential gene expression in AKP mouse intestinal cells upon exposure to atorvastatin, a drug used in the treatment of high cholesterol. For this experimental design, two replicates were used for both the control group and the atorvastatin-treated group. Like *DESeq2*, our input files will be separate count files generated by *featurecounts*. We will label our one factor 'Atorvastatin treatment' and denote our two groups as 'Treated' and 'Control', respectively. The corresponding *featurecounts* for our control and treated samples will be inputted to their respective groups. We must then specify the desired primary contrast, which dictates the exact comparison between samples that we are assessing. For example, if 'Treated-Control' is selected, this compares the gene expression of the treated samples with that of the control samples. If the order is switched and we instead write 'Control-Treated', the expression profile of the control samples will be compared to the treated samples. Although the genes identified will be identical regardless of the order of the terms in the '*contrast*' parameter, the classification of a gene as either upregulated or downregulated will differ based on the exact contrast being performed.

Under the 'Advanced options' setting, several parameters can be modified including the minimum LFC reported (default=0), the *p*-value cutoff for significance (default=0.05), and the method of *p*-value adjustment (default=Benjamini Hochberg). While these parameters were left to their default to generate the sample output (Figures 3.12 and 3.13), some parameters such as the *p*-value cutoff and minimum LFC could be modified, particularly if the output file contains a large number of DE genes between samples. The parameters used to generate the sample output are presented in Figure 3.14.

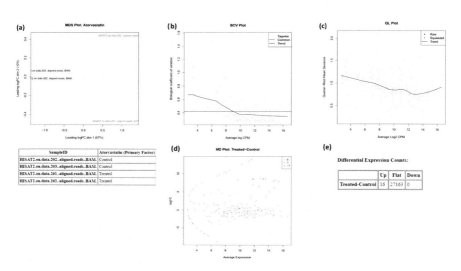

FIGURE 3.12 The graphical outputs generated by edgeR are presented. In order, the outputs include a plot showing a dimensionality reduction method that is very similar to PCA known as multidimensional scaling (MDS) (a), a plot showing the tagwise and common BCV calculations (b), a plot showing the tagwise and common QL calculations (c), a plot that shows the mean difference of the log-intensity ratios vs. the log-intensity averages (d), and a table showing the number of DE genes (e).

Normalization and Downstream Analyses

GeneID	logFC	logCPM	F	PValue	FDR
17075	8.56126890838074	13.5429015770018	529.962861204389	8.16662604172327e-07	0.00919874978074181
16691	8.61864116042631	13.2483584603569	516.046183016631	1.01481804851973e-06	0.00919874978074181
14319	8.5447209292265	13.1500704299152	515.948139995878	1.01535190191786e-06	0.00919874978074181
14181	6.48663644240555	12.6461591294218	432.658043619702	1.44224337854769e-06	0.00979968319638691
17110	7.93982581562463	14.6715355669555	382.0759019664	2.32889110941402e-06	0.0126593862925527
230163	8.46249819360597	11.8203528422206	309.054832637367	4.1775626709963e-06	0.0189236626391681
227613	5.06792970545537	14.05793858519235	241.572885914147	7.32874144024434e-06	0.028455409086343
66953	13.8847184050849	11.5558775298653	405.069357395914	1.02970946583472e-05	0.0323871858725406
17448	8.58746404043871	11.8205879032802	219.150115891751	1.07246283105657e-05	0.0323871858725406
12858	8.22844143091893	14.8952624927171	201.729783510611	1.34469010207568e-05	0.0337023035464625
20250	6.96731328801875	11.8242423641582	193.585634634769	1.50462911426245e-05	0.0337023035464625
100041948	8.21405303185871	9.73994141313242	190.869425704794	1.5637023181819e-05	0.0337023035464625
231507	8.73326167851449	12.7894191558013	188.749469303951	1.61201643218666e-05	0.0337023035464625
18458	7.83947870237615	15.388495495964	164.310009389484	2.35029774143065e-05	0.0440909608164988
20731	6.17991433313427	14.7513446320495	158.603373809018	2.58682201541635e-05	0.0440909608164988
105559	12.5942852085581	10.246373960767	286.865210983639	2.5955898784502e-05	0.0440909608164988

FIGURE 3.13 EdgeR gene list output for AKP treated vs. untreated.

3.4.3 BALLGOWN

The previously mentioned platforms *DESeq2* and edgeR require raw count data matrices that have not yet been normalized. In contrast, Ballgown utilizes pre-normalized count data (FPKM/RPKM) derived from Stringtie to assess differential gene expression across samples. If both data formats are available, the preferential tool to use is based upon the primary goal of the experimental design. If the purpose of the experiment is to simply identify DE genes for downstream pathway analysis, which will be discussed in the next section, the HISAT2-*featurecounts*-edgeR (*DESeq2*) pipeline is preferential. However, if the purpose is to instead study differential transcript usage as well as to identify novel transcripts, which is usually reserved for well-characterized species with detailed reference genomes available, the HISAT2–Stringtie–Ballgown is the best currently available pipeline. Regardless of the pipeline used however, the output files and graphs will be comprised of the same type of output including a list of DE genes, PCA plots, dispersion estimate plots, and heatmaps. Thus, the largest difference between these three platforms is the underlying statistical methods used to determine genes that are differentially expressed. Currently, Ballgown is not available as a tool in the usegalaxy.com interface so the subsequent pathway analysis described in the next section will be based on output files generated by *DESeq2*.

Stringtie ultimately generates three output tables that can be utilized by Ballgown that are labeled:

1. **Phenotype data:** This table contains sample-specific information such as treatment status.
2. **Expression data:** This table contains both raw and normalized abundances of exons, genes, and transcripts.
3. **Genomic data:** Contains genomic positions and annotations for the genomic features presented in the expression data table.

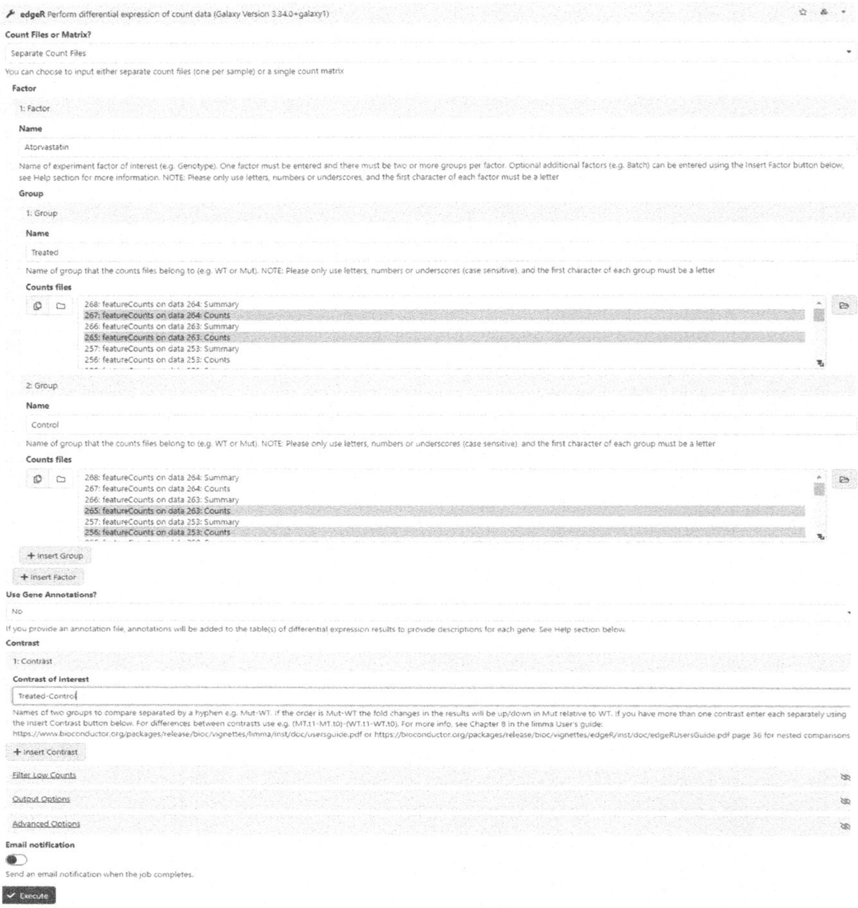

80   RNA-seq in Drug Discovery and Development

FIGURE 3.14   EdgeR parameters for AKP treated vs. untreated.

Unlike edgeR and *DESeq2*, Ballgown fits the abundance estimates (FPKM/RPKM) using standard linear models as opposed to GLMs. One possible consequence of using standard linear models is that the FPKM values can be highly skewed and have large variances. To remove this possible form of bias, the built-in functions in Ballgown stabilize the variance by applying a log transformation to the FPKM counts prior to fitting them with linear models. Like *DESeq2* and edgeR, the primary output for Ballgown is a series of plots including heatmaps and PCA plots as well as a list of DE genes containing fold changes and adjusted *p*-values.

Normalization and Downstream Analyses

3.5 VISUALIZATION OF DIFFERENTIAL EXPRESSION

Now that we have generated a list of DE genes, the final steps of RNA-seq experiments include the use of various visualization tools and pathway analysis browsers to determine underlying mechanistic drivers of the observed results as well as improve the interpretability of the results. Although there are many different plots and figures that can be used to present differential expression results, the three most common includes heatmaps, volcano plots, and sashimi plots. In addition, there are multiple web browsers that can be utilized to visualize and peruse complex genomic data including the UCSC genome browser previously mentioned as well as the Integrative Genomics Viewer (IGV). Finally, the following section will also discuss the process of pathway analysis using the bioinformatics tool Database for Annotation, Visualization, and Integrated Discovery (DAVID).

3.5.1 INTEGRATIVE GENOMICS VIEWER

The Integrative Genomics Viewer, which began development in 2007, is a high-performance, interactive visualization tool that can be used to view large heterogeneous datasets derived from RNA-seq experiments. IGV can either be directly utilized as a web browser by visiting the link https://igv.org/app/ or can be downloaded directly to your computer to use as a local file by visiting www.broadinstitute.org/igv. For ease of use, the following section will utilize the downloaded local version of the IGV tool.

Once the IGV tool is installed and open, the primary IGV application window will appear, in which there are multiple areas of interest as highlighted in Figure 3.15. Under the *file* tab in the top-left corner are options to upload datasets, primarily fastq files or .bam files (aligned reads), from local directories, dropbox, or from a direct web link URL.

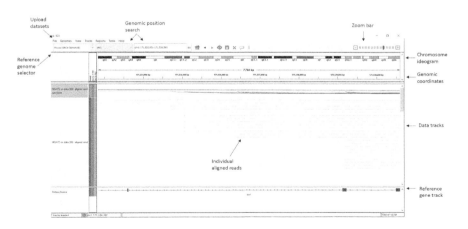

FIGURE 3.15 Labeled IGV application window.

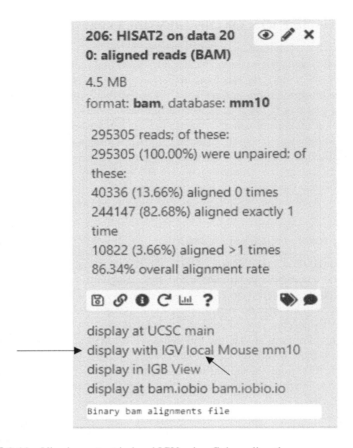

FIGURE 3.16 Viewing output in local IGV using Galaxy directly.

In Galaxy, a generated output file can be directly viewed by first selecting the output file of interest in the history sidebar and then clicking *display with IGV local* (Figure 3.16). Just below the *file* tab is a tab to select the reference genome. In most instances, the reference genome will be automatically detected once the data input is loaded into the IGV tool. To the right of the *reference genome* tab are search bars in which individuals can navigate and explore particular regions of the reference genome. This can be performed by either simply selecting the chromosome of interest from the first dropdown menu or by putting a specific chromosome coordinate into the adjacent search bar. For example, we can input chr1:1–400 to isolate the particular region of interest in chromosome 1. The name of the chromosome and the genomic positions must be separated by a colon in order to perform a successful search. In the top-right corner is the *zoom* function, which is necessary to see specific data points due to the vast size difference

Normalization and Downstream Analyses

between the entire genome and our individual's mapped reads. To view individual alignments, we must zoom to a level that is appropriate for the size of our reads (usually 5–10 kb is sufficient). If we zoom beyond this alignment visibility threshold, we are able to observe the specific amino acid sequence of the reads as well as the amino acid composition of the reference gene sequences.

IGV utilizes a color-coding scheme to denote areas of interest such as single nucleotide polymorphisms (SNPs). For example, at high zoom, areas in the reference gene that are colored red denote deletion events while green denotes substitution events. Below the *zoom* bar is an ideogram of the current chromosome selected that illustrates the current position being probed, which lies directly above a horizontal ruler that shows the exact coordinate a particular feature is found. The rest of the application window is occupied by our input data in the form of tracks, the names of which appear on the far-left side of the window. Tracks are simply horizontal data panels that present one sample or genomic annotation. Each gray line in the data track corresponds to a separate mapped read at that particular coordinate of the genome (Figure 3.15). IGV is able to view multiple genomic inputs simultaneously and will generate a separate track for each genomic file uploaded into the tool. Finally, the two tracks on the very bottom of the application window are populated by the reference genome sequences and the corresponding gene name annotations.

An additional useful visualization plot that can be generated through the IGV window is a sashimi plot. When provided with a gene annotation track (reference genome) and an aligned reads bam file as input, sashimi plots can be utilized to analyze splice junctions, assess differential exon usage, and identify splicing variants (alternative splicing events). For a given region of the genome, a sashimi plot will generate a visual plot governed by the following set of rules:

1. Exon alignments are represented as read densities. These read densities may be normalized by either length of the genomic region or coverage.
2. Reads that span splice junctions are presented as arcs that connect two exons together. The width of each arc is proportional to the number of reads that map to the splice junction.

In order to generate a sashimi plot in the IGV application window (Figure 3.17), the following steps can be followed:

1. Zoom out far enough so that the genomic region of interest is completely visible in the window.
2. Right click on the alignment track corresponding to the data of interest and select the *sashimi plot* option.
3. Select the feature track containing the annotation. By default, IGV will use the RefSeq genes track that is preloaded into IGV when the reference genome is selected. If more than one is available, IGV will prompt you to select which one will be utilized to generate the sashimi plot.

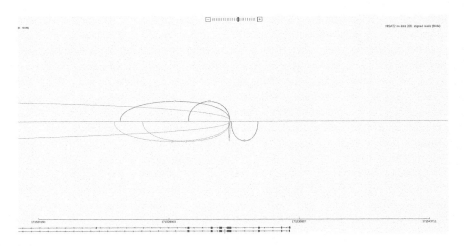

FIGURE 3.17 Sample sashimi plot generated in IGV. Blue-blocked areas at the bottom of the figure correspond to individual exons.

3.5.2 UCSC GENOME BROWSER

Like IGV, the UCSC Genome Browser is a publicly available database that can be used to visualize genomic sequence data and is integrated with many reference annotations for multiple species. The genome browser can be accessed through either the web link http://genome.ucsc.edu/ or directly through Galaxy by first clicking on the dataset of interest in the history sidebar and then clicking the *display at UCSC main* option. The UCSC Genome Browser contains positional tables for data based on start-stop coordinates and is defined using half-open zero-based ranges. For example, the first 100 bases correspond to a coordinate of *0,100* while bases 500–600 would be represented by the coordinate *500,600*. Once the dataset of interest has been loaded into the UCSC browser by the link embedded in Galaxy, the UCSC Genome Browser will generate a series of annotation tracks including mRNA sequencing, cross-species homologies, SNP data, gene predictions, expressed sequence tags, expression data, and others.

The main application window of the UCSC Genome Browser is presented in Figure 3.18. The topmost portion of the application window allows us to control how far the window is zoomed in or out. Below the *zoom in* and *zoom out* options is a search bar, which can make queries based on genomic coordinates or gene names. In the following example, we will use the name of the gene identified to be most differentially expressed between samples, interlectin 1 (*itln1*), as our query search. The UCSC browser will generate the aforementioned tracks for our sample, which we can access by clicking them directly. The tracks that are shown can be customized by scrolling to the bottom of the page and modifying the display properties under each category. For our purposes, the most useful track to click

Normalization and Downstream Analyses 85

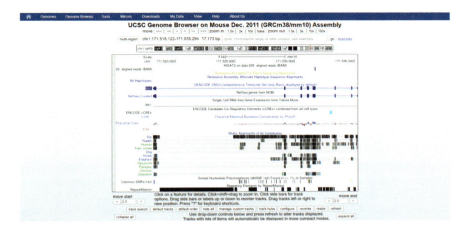

FIGURE 3.18 Main application window for the UCSC Genome Browser using the gene *itln1* as the search query.

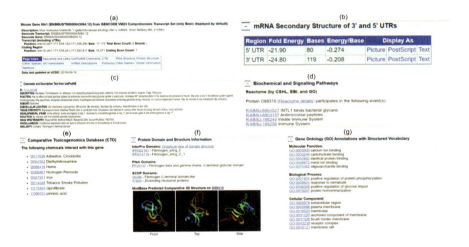

FIGURE 3.19 Comprehensive data for the *itln1* gene including gene names (a), mRNA secondary structures (b), gene description (c), biochemical signaling pathways (d), comparative toxicogenomics (e), protein structure (f), and gene ontology annotation (g).

on is the gene track itself. Clicking on this track will redirect to a page containing a plethora of gene-specific information including exon count, coding region coordinates, function, localization, comparative toxicogenomics, mRNA secondary structures, protein structures, gene ontology annotations, and biological signaling pathways that the gene of interest have been implicated in. Figure 3.19 shows samples of the aforementioned data for the *itln1* gene.

3.5.3 Heatmaps

Heatmaps are a prominent visualization tool that show differences in the numerical magnitude of a particular feature using color gradients. In the case of genomic data, color gradients are used to classify and cluster genes that are differentially expressed, whether overexpressed or underexpressed, across samples based on changes in LFC. In clustered heat maps, gene features are labeled horizontally while categories/samples are labeled vertically across the plot. Each individual cell corresponds to a discrete genomic feature in a given sample and the order of the vertical and horizontal labels can be adjusted accordingly.

Heatmaps are generated based on a hierarchical clustering method, which groups similar objects together and maximizes the distance between the clusters so that objects within one cluster are very similar to one another yet very distinct from components of a neighboring cluster. The process of hierarchical clustering initially treats all objects (genes in our case) as unique clusters and merges the clusters closest to one another. This process is then repeated until all clusters are merged together. The primary input for this process is a raw counts table and the corresponding package to generate heatmaps, which includes such tools as plotHeatmap, ggplot, and heatmap2, will also generate a distance matrix. A distance matrix computes the distance, referred to as the Euclidean distance, between each combination of objects in the input table. These computed Euclidean distances can then be used to construct a dendrogram, a figure in which expresses the distance between two objects based on the length of the line used to connect the two clusters. A sample dendrogram is presented in Figure 3.20.

To generate a heatmap figure in Galaxy, we will require two files: our original output file from *DESeq2* and a normalized count data table. As an overview, we will filter and extract our significant genes from the two aforementioned files and then combine them into one file for our heatmap input. The following steps, which

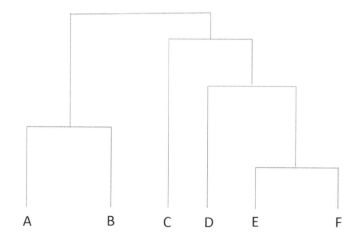

FIGURE 3.20 Sample dendrogram.

Normalization and Downstream Analyses

FIGURE 3.21 Parameters used for the *Filter data on any column using simple expressions* tool to filter and extract significant genes from the *DESeq2* output table.

are presented in Figure 3.21, must be conducted in order to filter out our significant DE genes from our *DESeq2* output:

1. Select the Filter data on any column using a simple expressions tool in Galaxy.
2. Set the Filter option to our *DESeq2* output file.
3. Set with the following condition option to c8>0.1. This will generate a new file containing only the genes that satisfy the given condition. In this case, column 8 corresponds to the adjusted *p*-value column and we are filtering out any genes with a *p*-value greater than 0.1.
4. Set the Number of header lines to skip to one. This will tell the tool to skip the first line in our input table, which corresponds to our column labels, and apply the above condition to all rows after the first.

Next, we will rerun *DESeq2* with the same job parameters stated above. However, this time we will select the output *rlog normalized table* option under the output options section. Once this job is completed, we will have a normalized count data table for each sample, which we will then merge with our original *DESeq2* output file. The following steps (Figure 3.22) will be used to combine our two files together:

1. Select the Join two Datasets side by side on a specific field tool.
2. Set the Join option to our original *DESeq2* output.
3. Set using column option to column 1. This corresponds to our gene ID column.
4. Set with option to our normalized count file.
5. Set the column option to column 1.
6. Set the Keep the header lines option to yes. Running this tool will generate a joined dataset containing all columns from both.

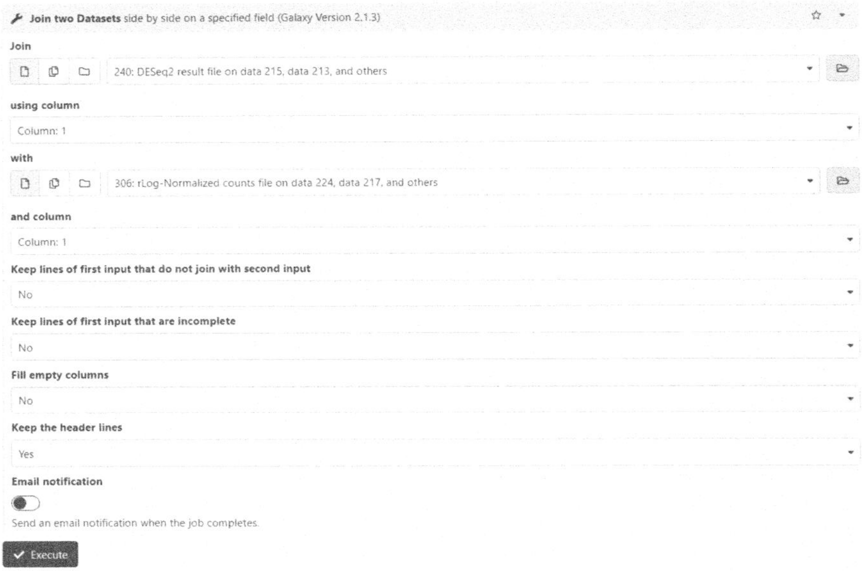

FIGURE 3.22 Parameters used for the *join two Datasets side by side on a specific field* tool to combine the *DESeq2* output table and the normalized counts table.

For the purposes of heatmap generation, this new merged file has some extraneous columns that are not required. We can use the *Cut columns from a table* tool to isolate our columns of interest in a brand new 'cut' file output that will populate our history. The following steps can be used to generate a new file (Figure 3.23) containing only the columns required to generate a heatmap:

1. Select the *Cut columns from a table* tool.
2. Set *cut columns* option to the columns that need to be isolated. For the purposes of the sample heatmap below, we only need the gene id column and the normalized counts for each sample individual. With this in mind, we will set this option to c8:c12.
3. Set the *delimited by* option to tab.
4. Set *from* option to the output from the *Join two Datasets side by side on a specific field* tool. This will generate a new file only containing the column selected in Step 2.

Now that we have generated our output file containing only the relevant columns required for the heatmap tool, the next step is to use the *heatmap2* tool available in Galaxy to produce our heatmap for the most significant DE genes. The following steps (Figure 3.24) were used to generate the sample heatmap presented in Figure 3.25:

Normalization and Downstream Analyses

FIGURE 3.23 Parameters used for the Cut columns *from a table* tool in order to isolate the columns required to generate a heatmap.

FIGURE 3.24 Parameters used to generate a sample heatmap using the *heatmap2* tool in Galaxy.

1. Select the *heatmap2* tool.
2. Set *input should have column headers* option to the output from the *cut columns from a table* tool.
3. Set the *data transformation* option to plot the data as is.
4. Set the *enable data clustering* option to no.
5. Set the *labeling columns and rows* option to label my columns and rows.
6. Set the *coloring groups* option to blue to white to red.
7. Set the *data scaling* option to *scale my data by row*.

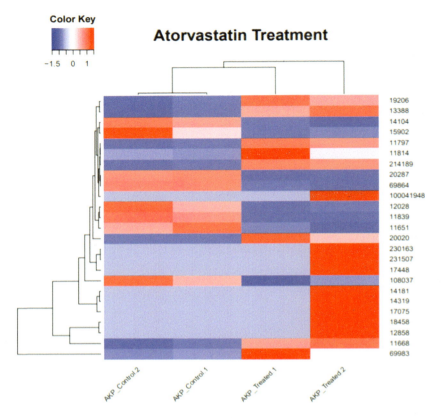

FIGURE 3.25 Heatmap comparing gene expression between atorvastatin-treated intestinal cells and untreated cells. The dendrogram is also included.

3.5.4 VOLCANO PLOTS

Volcano plots are a frequently used figure to present omics data by plotting statistical significance against magnitude of the change in the variable we are observing. In this case, volcano plots are a variant of scatterplots that plot the $\log_{10}$ of the adjusted *p*-value on the *y* axis against the log of the fold change in gene expression across two samples. Because of this log transformation of the *p*-values, lower *p*-values, which denote greater statistical significance, are located higher on the graph while higher *p*-values cluster further down the plot. In addition, red nodes represent upregulated genes while blue nodes denote downregulated genes. The characteristic two-arm shape of the plot, from which the name is derived, is due to a difference in assumed probability distribution for the *y* and *x* axes. The LFCs plotted on the *x* axis are assumed to be approximately normally distributed while the log *p*-values plotted on the *y* axis are skewed in such a way that higher changes in expression are associated with more significant *p*-values (lower values).

Normalization and Downstream Analyses

FIGURE 3.26 Parameters used to generate a sample volcano plot.

Although many packages exist that can generate volcano plots, we will use the *ggplot2*-based tool volcano plot tool in Galaxy. Our input file will be one of the tabular outputs containing our list of DE genes that were generated using either edgeR or *DESeq2*. We must then denote which columns in our input file corresponding to the adjusted *p*-value, the raw *p*-value, the LFC, and our gene ids. In the example below which utilizes the tabular output from *DESeq2*, the aforementioned sections correspond to column 7, column 6, column 3, and column 1, respectively. Next, we can set our level of statistical significance (default = 0.05) as well as select which points on the plot are to be labeled with their corresponding gene ids. In some cases, only a small number of genes may achieve a significance level of 0.01 or 0.05, which may inhibit further pathway analysis as too few genes are detected. In this case, it is justifiable to increase the level of significance to 0.1 or 0.15. Conversely, a large list of statistically significant genes may be further filtered by lowering the level of significance. Finally, additional parameters including plot title, axis labels, minimum and maximum values for the axes, and legend labels can be modified under the *plot options* tab. Figure 3.26 presents the parameters used to generate the sample volcano plot in Galaxy while the sample output is presented in Figure 3.27.

3.6 PATHWAY ANALYSIS USING THE NIH DAVID WEB SERVER

The final step in many differential expression analysis experiments is determining significantly dysregulated molecular mechanisms based on the list of DE genes. For the example below, the free source DAVID will be used to map statistically significant genes to molecular pathways. The DAVID tool can be accessed by using the web link https://david.ncifcrf.gov/tools.jsp. The following steps,

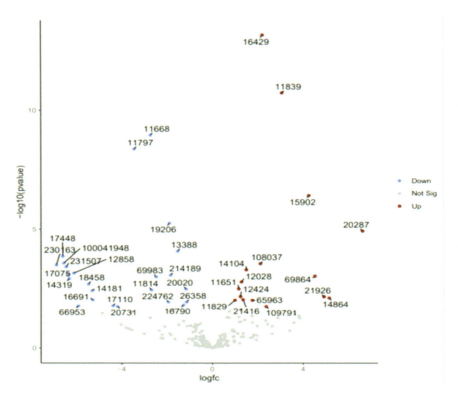

FIGURE 3.27 Sample volcano plot is presented. Significantly upregulated genes are labeled with red nodes and significantly downregulated genes are labeled with blue nodes. Non-DE genes are denoted by gray nodes.

summarized in Figures 3.28–3.30, will be used to generate molecular pathway figures based on a list of significant DE genes:

1. Copy and paste a list of significantly differentiated genes. As an alternative, a local file may be uploaded instead of simply copying and pasting (Figure 3.28).
2. Select the gene identifier, i.e., the source of the gene id. In the above example, the gene ids are entrez_gene_ids.
3. Select the gene list under the *list type* option.
4. Submit the list.
5. Select the species that the genes belong to (Figure 3.29).
6. Select *Pathways*.
7. Select *KEGG_Pathway* (Figure 3.30). The Kyoto Encyclopedia of Genes and Genomes (KEGG) is a public database containing extensive information concerning gene functionality and pathways.

Normalization and Downstream Analyses

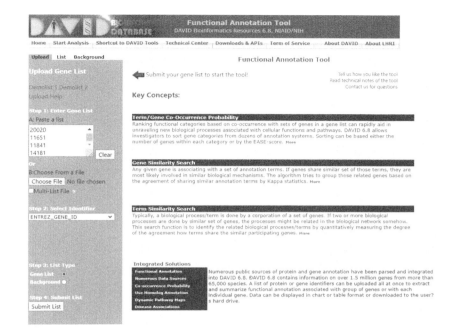

FIGURE 3.28 Enter a list of significantly differentiated genes in the NIH DAVID web server.

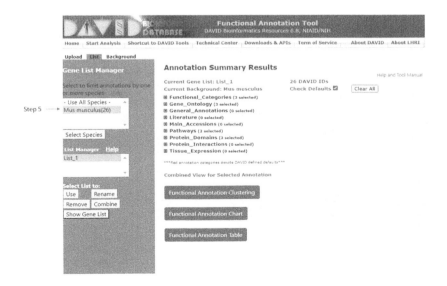

FIGURE 3.29 Select the species that the genes belong to in the NIH DAVID web server.

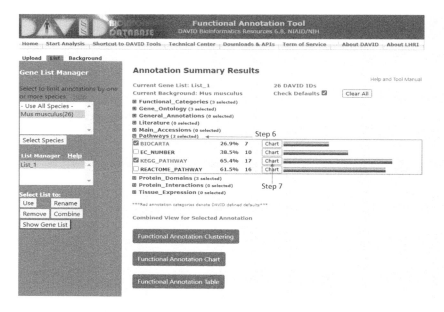

FIGURE 3.30 Select KEGG_Pathway in the NIH DAVID web server.

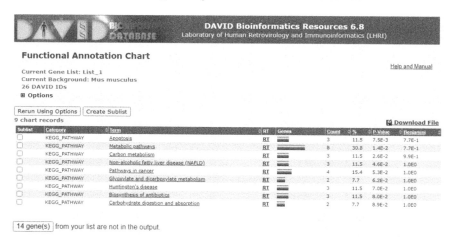

FIGURE 3.31 Annotated pathways based on a list of significant DE genes in atorvastatin-treated intestinal cells in AKP mice. Note that no pathway achieved statistical significance.

Once we chart the *KEGG_Pathways*, we then generate a list of annotated molecular mechanisms in which significant genes on the submitted list map to along with associated *p*-values for each pathway in a separate window. As evident by the output (Figure 3.31), no statistically significant molecular mechanisms were identified.

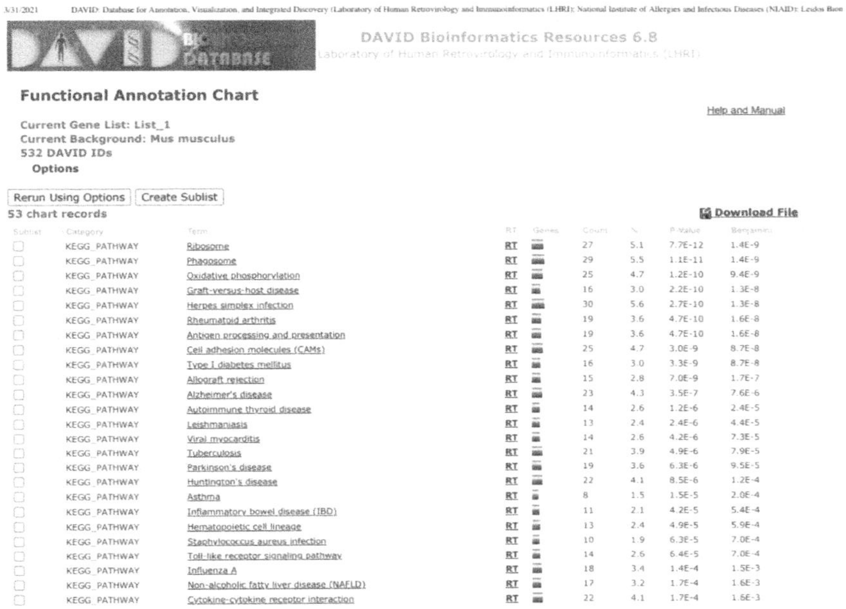

3/31/2021 DAVID: Database for Annotation, Visualization, and Integrated Discovery (Laboratory of Human Retrovirology and Immunoinformatics (LHRI); National Institute of Allergies and Infectious Diseases (NIAID); Leidos Bion

DAVID Bioinformatics Resources 6.8
Laboratory of Human Retrovirology and Immunoinformatics (LHRI)

Functional Annotation Chart

Help and Manual

Current Gene List: List_1
Current Background: Mus musculus
532 DAVID IDs
Options

Rerun Using Options | Create Sublist

53 chart records

Download File

Sublist	Category	Term	RT	Genes	Count	%	P-Value	Benjamini
	KEGG_PATHWAY	Ribosome	RT		27	5.1	7.7E-12	1.4E-9
	KEGG_PATHWAY	Phagosome	RT		29	5.5	1.1E-11	1.4E-9
	KEGG_PATHWAY	Oxidative phosphorylation	RT		25	4.7	1.2E-10	9.4E-9
	KEGG_PATHWAY	Graft-versus-host disease	RT		16	3.0	2.2E-10	1.3E-8
	KEGG_PATHWAY	Herpes simplex infection	RT		30	5.6	2.7E-10	1.3E-8
	KEGG_PATHWAY	Rheumatoid arthritis	RT		19	3.6	4.7E-10	1.6E-8
	KEGG_PATHWAY	Antigen processing and presentation	RT		19	3.6	4.7E-10	1.6E-8
	KEGG_PATHWAY	Cell adhesion molecules (CAMs)	RT		25	4.7	3.0E-9	8.7E-8
	KEGG_PATHWAY	Type I diabetes mellitus	RT		16	3.0	3.3E-9	8.7E-8
	KEGG_PATHWAY	Allograft rejection	RT		15	2.8	7.0E-9	1.7E-7
	KEGG_PATHWAY	Alzheimer's disease	RT		23	4.3	3.5E-7	7.6E-6
	KEGG_PATHWAY	Autoimmune thyroid disease	RT		14	2.6	1.2E-6	2.4E-5
	KEGG_PATHWAY	Leishmaniasis	RT		13	2.4	2.4E-6	4.4E-5
	KEGG_PATHWAY	Viral myocarditis	RT		14	2.6	4.2E-6	7.3E-5
	KEGG_PATHWAY	Tuberculosis	RT		21	3.9	4.9E-6	7.9E-5
	KEGG_PATHWAY	Parkinson's disease	RT		19	3.6	6.3E-6	9.5E-5
	KEGG_PATHWAY	Huntington's disease	RT		22	4.1	8.5E-6	1.2E-4
	KEGG_PATHWAY	Asthma	RT		8	1.5	1.5E-5	2.0E-4
	KEGG_PATHWAY	Inflammatory bowel disease (IBD)	RT		11	2.1	4.2E-5	5.4E-4
	KEGG_PATHWAY	Hematopoietic cell lineage	RT		13	2.4	4.9E-5	5.9E-4
	KEGG_PATHWAY	Staphylococcus aureus infection	RT		10	1.9	6.3E-5	7.0E-4
	KEGG_PATHWAY	Toll-like receptor signaling pathway	RT		14	2.6	6.4E-5	7.0E-4
	KEGG_PATHWAY	Influenza A	RT		18	3.4	1.4E-4	1.5E-3
	KEGG_PATHWAY	Non-alcoholic fatty liver disease (NAFLD)	RT		17	3.2	1.7E-4	1.6E-3
	KEGG_PATHWAY	Cytokine-cytokine receptor interaction	RT		22	4.1	1.7E-4	1.6E-3

FIGURE 3.32 Sample KEGG pathway analysis containing a list of statistically significant pathways that are dysregulated.

This suggests that no molecular pathways were significantly dysregulated in mouse intestinal cells treated with atorvastatin. In Figure 3.32, we present a list of annotated molecular pathways that contain pathways that achieve statistical significance using a different gene list. Figure 3.33 presents a sample KEGG pathway generated for the carbon metabolism pathway.

3.7 SUMMARY

In this chapter, we initially discussed how to quantify our transcript expression as well as how to extract raw counts using Stringtie and *featurecounts,* respectively. Next, we discussed the primary methods in which counts are normalized, particularly based on library size using RPKM and FPKM for single-end and paired-end reads, respectively. Following that, we discussed three possible tools to assess differential gene expression across samples including *DESeq2* and edgeR, which usually follow *featurecounts* in a RNA-seq experiment pipeline, and the *r* package Ballgown, which is usually performed after use of Stringtie, as well as how to utilize them in Galaxy. Despite yielding similar results, the underlying differences in statistical workings of the aforementioned differential expression tools were also discussed. Finally, this chapter discussed common data representation formats including heatmaps, volcano plots, and sashimi plots as well as publicly

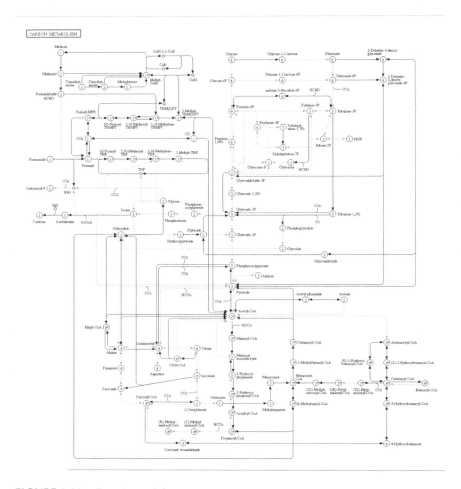

FIGURE 3.33 Sample KEGG pathway graphical representation.

available genome browsers such as the UCSC Genome Browser and IGV. Pathway analysis was explored using a list of significant DE genes as input for DAVID in order to find significantly dysregulated pathways in the atorvastatin treatment experiment.

KEYWORDS AND PHRASES

After reading this chapter, you should be able to demonstrate familiarity with the following words and phrases:

- Understand how the Stringtie algorithm quantifies transcript abundance.
- Understand the difference between Stringtie and tools that generate raw count matrices.

Normalization and Downstream Analyses

- Understand how featurecounts works and how to run it successfully in Galaxy.
- Be able to describe the three major ways in which libraries are normalized including (1) by library size, (2) by distribution/testing, and (3) by controls.
- Be able to calculate TPM, RPKM, and FPKM.
- Be able to understand and explain the idea of differential expression analysis.
- Be able to differentiate between which tools require count files generated by Stringtie (Ballgown) and those that require raw count matrices (DESeq and edgeR).
- Understand and be able to effectively utilize the UCSC and IGV genome browsers.
- Be able to interpret heatmaps and volcano plots as well as how to successfully generate them in Galaxy.
- Understand what pathway enrichment analysis is and how to perform it using DAVID.

BIBLIOGRAPHY

Anders, S., & Huber, W. (2010). Differential expression analysis for sequence count data. *Genome Biol, 11*(10), R106. https://doi.org/10.1186/gb-2010-11-10-r106

Anders, S., Pyl, P. T., & Huber, W. (2015, Jan 15). HTSeq: A python framework to work with high-throughput sequencing data. *Bioinformatics, 31*(2), 166–169. https://doi.org/10.1093/bioinformatics/btu638

Andrecut, M. (2009, Nov). Parallel GPU implementation of iterative PCA algorithms. *J Comput Biol, 16*(11), 1593–1599. https://doi.org/10.1089/cmb.2008.0221

Batut, B., Hiltemann, S., Bagnacani, A., Baker, D., Bhardwaj, V., Blank, C., Bretaudeau, A., Brillet-Gueguen, L., Cech, M., Chilton, J., Clements, D., Doppelt-Azeroual, O., Erxleben, A., Freeberg, M. A., Gladman, S., Hoogstrate, Y., Hotz, H. R., Houwaart, T., Jagtap, P., Lariviere, D., Le Corguille, G., Manke, T., Mareuil, F., Ramirez, F., Ryan, D., Sigloch, F. C., Soranzo, N., Wolff, J., Videm, P., Wolfien, M., Wubuli, A., Yusuf, D., Galaxy Training, N., Taylor, J., Backofen, R., Nekrutenko, A., & Gruning, B. (2018, Jun 27). Community-driven data analysis training for biology. *Cell Syst, 6*(6), 752–758. https://doi.org/10.1016/j.cels.2018.05.012

Broom, B. M., Ryan, M. C., Brown, R. E., Ikeda, F., Stucky, M., Kane, D. W., Melott, J., Wakefield, C., Casasent, T. D., Akbani, R., & Weinstein, J. N. (2017, Nov 1). A galaxy implementation of next-generation clustered heatmaps for interactive exploration of molecular profiling data. *Cancer Res, 77*(21), e23–e26. https://doi.org/10.1158/0008-5472.CAN-17-0318

Conesa, A., Madrigal, P., Tarazona, S., Gomez-Cabrero, D., Cervera, A., McPherson, A., Szczesniak, M. W., Gaffney, D. J., Elo, L. L., Zhang, X., & Mortazavi, A. (2016, Jan 26). A survey of best practices for RNA-seq data analysis. *Genome Biol, 17*, 13. https://doi.org/10.1186/s13059-016-0881-8

Evans, C., Hardin, J., & Stoebel, D. M. (2018, Sep 28). Selecting between-sample RNA-Seq normalization methods from the perspective of their assumptions. *Brief Bioinform, 19*(5), 776–792. https://doi.org/10.1093/bib/bbx008

Federico, A., Serra, A., Ha, M. K., Kohonen, P., Choi, J. S., Liampa, I., Nymark, P., Sanabria, N., Cattelani, L., Fratello, M., Kinaret, P. A. S., Jagiello, K., Puzyn, T., Melagraki, G., Gulumian, M., Afantitis, A., Sarimveis, H., Yoon, T. H.,

Grafstrom, R., & Greco, D. (2020, May 8). Transcriptomics in toxicogenomics: Preprocessing and differential expression analysis for high quality data. *Nanomaterials (Basel), 10*(5), 41–46. https://doi.org/10.3390/nano10050903

Frazee, A. C., Pertea, G., Jaffe, A. E., Langmead, B., Salzberg, S. L., & Leek, J. T. (2015, Mar). Ballgown bridges the gap between transcriptome assembly and expression analysis. *Nat Biotechnol, 33*(3), 243–246. https://doi.org/10.1038/nbt.3172

Huang da, W., Sherman, B. T., & Lempicki, R. A. (2009a, Jan). Bioinformatics enrichment tools: Paths toward the comprehensive functional analysis of large gene lists. *Nucleic Acids Res, 37*(1), 1–13. https://doi.org/10.1093/nar/gkn923

Huang da, W., Sherman, B. T., & Lempicki, R. A. (2009b). Systematic and integrative analysis of large gene lists using DAVID bioinformatics resources. *Nat Protoc, 4*(1), 44–57. https://doi.org/10.1038/nprot.2008.211

Kaisers, W., Schwender, H., & Schaal, H. (2018, Nov 21). Hierarchical clustering of DNA *k*-mer Counts in RNAseq Fastq files identifies sample heterogeneities. *Int J Mol Sci, 19*(11), 32–37. https://doi.org/10.3390/ijms19113687

Kanehisa, M., Furumichi, M., Tanabe, M., Sato, Y., & Morishima, K. (2017, Jan 4). KEGG: New perspectives on genomes, pathways, diseases and drugs. *Nucleic Acids Res, 45*(D1), D353–D361. https://doi.org/10.1093/nar/gkw1092

Karolchik, D., Baertsch, R., Diekhans, M., Furey, T. S., Hinrichs, A., Lu, Y. T., Roskin, K. M., Schwartz, M., Sugnet, C. W., Thomas, D. J., Weber, R. J., Haussler, D., & Kent, W. J. (2003, Jan 1). The UCSC genome browser database. *Nucleic Acids Res, 31*(1), 51–54. https://doi.org/10.1093/nar/gkg129

Katz, Y., Wang, E. T., Silterra, J., Schwartz, S., Wong, B., Thorvaldsdottir, H., Robinson, J. T., Mesirov, J. P., Airoldi, E. M., & Burge, C. B. (2015, Jul 15). Quantitative visualization of alternative exon expression from RNA-seq data. *Bioinformatics, 31*(14), 2400–2402. https://doi.org/10.1093/bioinformatics/btv034

Kovaka, S., Zimin, A. V., Pertea, G. M., Razaghi, R., Salzberg, S. L., & Pertea, M. (2019, Dec 16). Transcriptome assembly from long-read RNA-seq alignments with StringTie2. *Genome Biol, 20*(1), 278. https://doi.org/10.1186/s13059-019-1910-1

Li, W. (2012, Dec). Volcano plots in analyzing differential expressions with mRNA microarrays. *J Bioinform Comput Biol, 10*(6), 1231003. https://doi.org/10.1142/S0219720012310038

Liao, Y., Smyth, G. K., & Shi, W. (2014, Apr 1). Counts: An efficient general purpose program for assigning sequence reads to genomic features. *Bioinformatics, 30*(7), 923–930. https://doi.org/10.1093/bioinformatics/btt656

Liao, Y., Smyth, G. K., & Shi, W. (2019, May 7). The R package Rsubread is easier, faster, cheaper and better for alignment and quantification of RNA sequencing reads. *Nucleic Acids Res, 47*(8), e47. https://doi.org/10.1093/nar/gkz114

Love, M. I., Huber, W., & Anders, S. (2014). Moderated estimation of fold change and dispersion for RNA-seq data with DESeq2. *Genome Biol, 15*(12), 550. https://doi.org/10.1186/s13059-014-0550-8

Navarro Gonzalez, J., Zweig, A. S., Speir, M. L., Schmelter, D., Rosenbloom, K. R., Raney, B. J., Powell, C. C., Nassar, L. R., Maulding, N. D., Lee, C. M., Lee, B. T., Hinrichs, A. S., Fyfe, A. C., Fernandes, J. D., Diekhans, M., Clawson, H., Casper, J., Benet-Pages, A., Barber, G. P., Haussler, D., Kuhn, R. M., Haeussler, M., & Kent, W. J. (2021, Jan 8). The UCSC genome browser database: 2021 Update. *Nucleic Acids Res, 49*(D1), D1046–D1057. https://doi.org/10.1093/nar/gkaa1070

Pertea, M., Kim, D., Pertea, G. M., Leek, J. T., & Salzberg, S. L. (2016, Sep). Transcript-level expression analysis of RNA-seq experiments with HISAT, StringTie and Ballgown. *Nat Protoc, 11*(9), 1650–1667. https://doi.org/10.1038/nprot.2016.095

Pertea, M., Pertea, G. M., Antonescu, C. M., Chang, T. C., Mendell, J. T., & Salzberg, S. L. (2015, Mar). StringTie enables improved reconstruction of a transcriptome from RNA-seq reads. *Nat Biotechnol, 33*(3), 290–295. https://doi.org/10.1038/nbt.3122

Robinson, M. D., McCarthy, D. J., & Smyth, G. K. (2010, Jan 1). edgeR: A bioconductor package for differential expression analysis of digital gene expression data. *Bioinformatics, 26*(1), 139–140. https://doi.org/10.1093/bioinformatics/btp616

Thorvaldsdottir, H., Robinson, J. T., & Mesirov, J. P. (2013, Mar). Integrative genomics viewer (IGV): High-performance genomics data visualization and exploration. *Brief Bioinform, 14*(2), 178–192. https://doi.org/10.1093/bib/bbs017

4 Constitutive and Alternative Splicing Events

Robert Morris and Feng Cheng
University of South Florida

CONTENTS

4.1 What Is Splicing? 101
4.2 Molecular Mechanism of Splicing 102
4.3 Alternative Splicing 104
4.4 Differential Splicing Analysis 106
    4.4.1 Cuffdiff 2 106
    4.4.2 DiffSplice..109
    4.4.3 DEXSeq 112
    4.4.4 edgeR 118
    4.4.5 LIMMA 123
4.5 Summary 123
Keywords and Phrases 124
Bibliography 126

4.1 WHAT IS SPLICING?

RNA splicing in eukaryotic cells is a molecular process in which nascent precursor mRNAs containing all introns and exons for a given gene, termed pre-mRNA, are processed and trimmed down to mature mRNA transcripts. In splicing, exons are segments of a DNA or RNA molecule containing coding information required for the synthesis of proteins while introns are noncoding sequences separating distinct exons that are typically removed from the mRNA molecule prior to the formation of a mature transcript and subsequent translation. RNA splicing is a process that is intimately linked to transcription and is responsible for the diverse number of proteins produced by eukaryotic genomes. RNA splicing can be classified as either constitutive splicing, which occurs when all introns are removed and each exon is ligated together, or alternative splicing, a mechanism of which generates multiple isoforms for each gene due to the generation of transcripts with differing numbers of included exons. Splicing events are mediated by the spliceosome, a large

DOI: 10.1201/9781003174028-4

ribonucleoprotein complex composed of both proteins and small nuclear RNAs, and occur almost exclusively in the nucleus of eukaryotic cells.

4.2 MOLECULAR MECHANISM OF SPLICING

Within each intron, there are three critical regions that are required for the assembly of the spliceosome as well as for the completion of successful splicing. The first region of interest is the 5′ donor site, which is located at the 5′ end of the intron near the intron/exon boundary and contains a highly conserved GU nucleotide sequence. The second region of interest, termed the 3′ acceptor site, is located at the 3′ end of the intron and contains a highly conserved AG nucleotide sequence. Finally, directly upstream of the polypyrimidine tract, a region of the mRNA with a high number of C and U nucleotides, lies the branchpoint. Within the branchpoint lies a significant A residue that is responsible for initiating the formation of the lariat structure and subsequent splicing events. Figure 4.1 illustrates the structure and positioning of the key splicing sites in a pre-mRNA molecule.

RNA splicing of pre-mRNA is mediated by the formation of the spliceosome, which occurs in a stepwise manner. The following steps occur during the canonical splicing of a pre-mRNA into a mature transcript (Figure 4.2):

1. U1 small nuclear ribonucleoprotein (snRNP) binds to the conserved GU nucleotide sequence in the 5′ splice site.
2. Splicing factor 1 binds to the branchpoint sequence in the intron.
3. The auxiliary factor U2AF binds to the polypyrimidine tract and 3′ splice site.
4. U2AF recruits the snRNP U2 to the branchpoint, resulting in the displacement of splicing factor 1 and the hydrolyzing of ATP.
5. The U5/U4/U6 snRNP trimer binds to the complex with U5 binding to the 5′ splice site and U6 binding to U2.
6. A conformational change occurs where U1 is released from the complex and U6 replaces U5 at the 5′ splice site.
7. U4 is released from the complex and U6/U2 catalyzes the two-step transesterification reaction that results in the mature mRNA transcript.

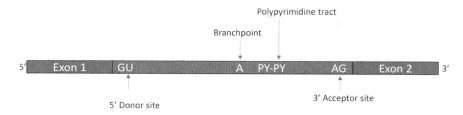

FIGURE 4.1 A portion of a pre-mRNA is presented. The locations of the 5′ donor site, 3′ acceptor site, polypyrimidine tract, and branchpoint within the intron are labeled.

Constitutive and Alternative Splicing Events

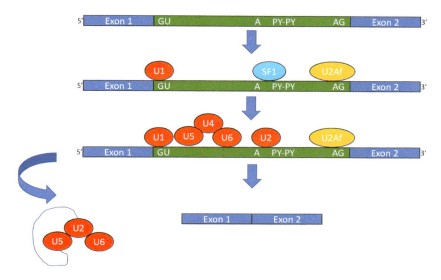

FIGURE 4.2 Overview of the molecular process of splicing is presented.

The transesterification reaction occurs in two sequential steps. In the first step, a 2′-OH group in the conserved A nucleotide located in the branchpoint performs a nucleophilic attack on the conserved GU nucleotide sequence of the 5′ splice site (Figure 4.3). This

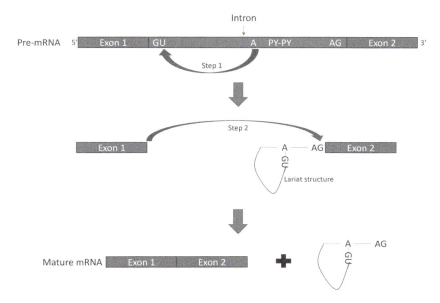

FIGURE 4.3 The two-step transesterification process that occurs during splicing is presented.

biochemical step results in the formation of the lariat structure, an intronic structure with a 2′–5′ phosphodiester bond and an overall circular shape. In the second step, a newly exposed 3′-OH group at the 5′ splice site then initiates a nucleophilic attack on the conserved AG sequence at the 3′ intron/exon boundary. This second nucleophilic attack ligates the two exons together and releases the intronic lariat.

4.3 ALTERNATIVE SPLICING

Alternative splicing utilizes the same principles of constitutive splicing to generate a large pool of distinct protein variants known as isoforms and is primarily responsible for the significantly larger number of unique protein species (>90,000 proteins in humans) relative to the number of genes (~25,000 genes). Alternative splicing may result in transcripts with fewer exons than transcripts produced by constitutive splicing or may include intronic sequences in the final mature mRNA product. Because these transcripts have differing nucleotide sequences, the resulting proteins will be composed of differing amino acid residues and as a result, will have unique properties and functions. In general, there are five types of alternative splicing events:

1. mutually exclusive exons;
2. exon skipping;
3. alternative 3′ splice site;
4. alternative 5′ splice site; and
5. intron retention.

Figure 4.4 shows the comparison between constitutive splicing and the five major variations of alternative splicing. In general, the most frequently occurring

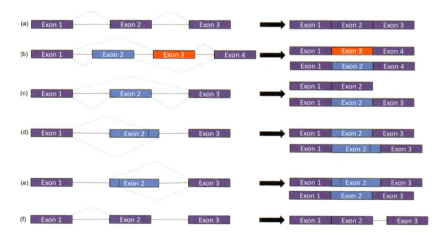

FIGURE 4.4 Sample diagrams for constitutive splicing (a), mutually exclusive exons (b), exon skipping (c), alternative 3′ splice site (d), alternative 5′ splice site (e), and intron retention (f) are presented.

Constitutive and Alternative Splicing Events

alternative splicing event in higher order vertebrates and invertebrates is exon skipping while intron retention is most common in lower metazoans. For vertebrates, intron retention typically occurs only in untranslated regions and is associated with weaker splice sites and shorter intron length. Splice sites may be classified as either weak or strong with weak splice sites usually associated with alternative splicing and strong splice sites associated with constitutive splicing. The strength of a given splice site is directly correlated with the frequency in which a splice site is recognized and used. The determination of splice site strength is largely dependent on the sequence that accompanies the highly conserved GU and AG nucleotides in 5' splice sites and 3' splice sites, respectively. Weak splice sites are particularly prevalent in exons that contain alternative 5' or 3' splice sites in which one of the variables' starting or ending positions is less favorable than the other.

The overall splicing efficiency for a given splice site is largely influenced by the presence of splicing regulatory elements, short sequence motifs that may either promote or hinder the assembly of the spliceosome at a particular splice site or may affect access to other regulatory elements. These motifs are classified based on whether or not they are located in intronic or exonic regions and whether or not they promote or inhibit splicing progression. This includes exonic splicing enhancers (ESEs), exonic splicing silencers, intronic splicing enhancers, and intronic splicing silencers. These cis-acting elements that are sequences located directly in the pre-mRNA then interact with and bind trans-acting RNA binding proteins to regulate splicing, particularly proteins belonging to the serine–arginine-rich (SR) proteins and heterogeneous nuclear ribonucleoprotein (hnRNP) families, in a sequence-dependent manner.

Figure 4.5a presents the promotion of splicing by the binding of SR proteins and the recruitment of spliceosome components. Figure 4.5b illustrates exon exclusion

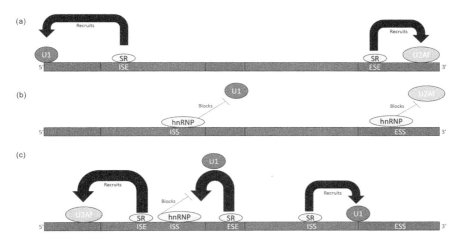

FIGURE 4.5 Examples for activation of splicing (a), inhibition of splicing (b), and a combination of both activator and repressor activity (c) are presented.

106 RNA-seq in Drug Discovery and Development

by blocking the recruitment of spliceosome components. Figure 4.5c demonstrates both principles in which splicing is promoted for one exon but excluded for the downstream exon. In general, SR proteins promote splicing at a given splice site by recruiting U1 and U2AF to their respective positions for proper spliceosome assembly and will bind either ESEs or intronic splicing enhancers, while hnRNP proteins will inhibit splicing by binding to exonic splicing silencers and intronic splicing silencer sequences. hnRNPs serve a repressive role in the modulation of alternative splicing by generating a loop structure for an adjacent exon that blocks the binding of pro-slicing SR proteins. Although SR proteins and hnRNPs are generally viewed as activators and inhibitors of splicing, respectively, the positioning and overall combination of these splicing regulators ultimately determine whether or not splicing occurs. For example, SRSF10 may promote exon inclusion and splicing when it binds to an ESE located in one exon, but will instead promote exon exclusion if it binds to an ESE sequence in the downstream exon. Additional layers of regulation for splicing arise from secondary structures in the pre-mRNA. For instance, a RNA hairpin may block a 5' or 3' splice site from being recognized by the spliceosome, resulting in exon exclusion.

4.4 DIFFERENTIAL SPLICING ANALYSIS

Multiple tools exist to assess differential splicing across conditions, many of which are coupled with analyses for differential gene expression. Differential splicing tools can be divided into one of two groups based upon the underlying principles they employ to detect alternative splicing events. The first group includes isoform-based programs such as Cuffdiff 2 and DiffSplice, which estimate the difference in the relative expression of gene transcripts and use these estimates to draw inferences concerning alterations in alternative splicing patterns. The second major approach for differential splicing analysis utilizes count-based methods and can further be subdivided into exon-based approaches (includes DEXSeq and the previously mentioned *limma* and edgeR) as well as event-based methods such as SUPPA2 and rMATS. In this section, we will primarily focus on the use of isoform-based and exon-based approaches as they pertain to the analysis of differential splicing.

4.4.1 CUFFDIFF 2

Cuffdiff 2, a component contained within the Cufflinks software suite, is an algorithm that estimates transcript-level expression profiles to assess differential splicing. Assuming we have set up a differential analysis study in which we are assessing differential isoform expression between at least two conditions, we can measure the expression of a particular transcript by counting the number of fragments that mapped to a particular region of the genome and then compare the fragment counts across samples. The traditional use of a Poisson model, which estimates variability as a function of the mean transcript counts across replicates, fails to account for two primary issues that arise when assessing differential splicing. The first complication, termed count uncertainty, occurs due to the

Constitutive and Alternative Splicing Events

observation that up to 50% of reads will map with low confidence. That is, many reads will map to multiple regions of the genome and thus we cannot decisively conclude the exact origin of the transcript. This occurs because most splice variants for most genes share large homologous sequences in higher eukaryotes. In addition, many genes have multiple copies within the genome, termed paralogs, that also have a high degree of similarity between them. The second complication that arises is count overdispersion, a phenomenon that frequently occurs when dealing with count data where the sample variability is higher than predicted by a traditional Poisson model.

To address these possible issues, Cuffdiff 2 models the variability of a transcript's fragment account in a manner that is dependent on both the expression profile and its splicing structure. Cuffdiff 2 first measures the overdispersion in the experimental conditions by fitting the observed variances, in fragment counts, as a function of the mean across samples. Next, the Cuffdiff algorithm estimates the number of fragments originating from each individual transcript and then combines the uncertainty in each fragment count with the overdispersion estimates predicted for that fragment as estimated by the previously generated global model of cross-replicate variability. Ultimately, the algorithm assumes a beta probability distribution for the fragment counts while assuming a negative binomial distribution for the overdispersion estimates. This combined probability distribution, termed a beta negative binomial distribution, refers to a distribution for a particular discrete random variable termed x that is equal to the number of failures required to reach a certain number of successes, termed r, in a series of independent Bernoulli trials (two possible outcomes). The probability of success for each trial, termed p, is assumed to be constant within any given experiment, but is treated as a random variable that adheres approximately to a beta distribution between different experimental conditions.

As input, Cuffdiff 2 requires a GTF file generated from either the transcript assembler Cufflinks or a related package as well as at least two alignment (.bam) files for two conditions. The output for a completed Cuffdiff 2 run in Galaxy generates the following 11 files:

1. Transcript FPKM expression tracking
2. **Gene FPKM expression tracking**: File that tracks summed FPKM for transcripts that share each gene_id.
3. **Primary transcript FPKM tracking**: File that tracks the summed FPKM of all transcripts that share each tss_id.
4. **Coding sequence FPKM tracking**: File that tracks the summed FPKM of all transcripts that share a p_id, regardless of whether or not they share a tss_id.
5. Transcript differential FPKM
6. **Gene differential FPKM**: Tests the differences in the summed FPKM of all transcripts that share a gene-id across samples.
7. **Primary transcript differential FPKM**: Tests the differences in the summed FPKM of all transcripts that share a tss_id across samples.

8. **Coding sequence differential FPKM**: Tests the differences in the summed FPKM of all transcripts that share a p_id, regardless of whether or not they share the same tss_id.
9. **Differential splicing tests**: A tab-delimited file that lists how much differential splicing between all isoforms derived from a single primary transcript.
10. **Differential promoter tests**: A tab-delimited file that lists how much differential promoter use is present across samples for each gene.
11. **Differential CDS tests**: A tab-delimited file that lists how much differential coding sequence (CDS) output is present across samples.

The traditional pipeline for differential expression, referred to as the 'Tuxedo' pipeline, utilizes the alignment tool TopHat, the transcript assembler Cufflinks, and the differential expression tool Cuffdiff 2. Figure 4.6 provides an overview of the traditional Tuxedo pipeline. First, raw reads in either .fasta or .fastq format are aligned to a reference genome using the TopHat tool. Then, full transcripts and possible splicing alternatives are assembled from the aligned reads output generated from TopHat through the Cufflinks tool, producing .gtf files containing estimates of transcript expression. Cuffmerge, a tool that is part of the Cufflinks suite of tools, can then be used to parsimoniously merge these annotated transcript files

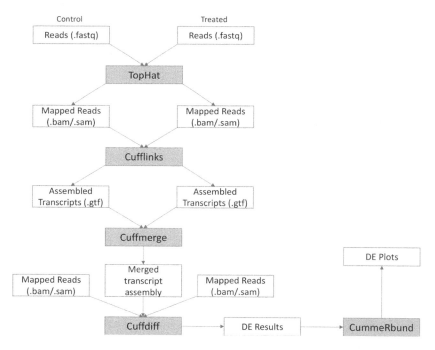

FIGURE 4.6 Outline of the traditional 'Tuxedo' pipeline that uses Cuffdiff to assess the differential expression and differential splicing events.

Constitutive and Alternative Splicing Events

with a reference genome in order to generate a merged.gtf file. Finally, Cuffdiff 2 can be used to calculate whether or not differential expression of isoforms is statistically significant across multiple samples. To simplify the analysis of Cuffdiff 2 outputs, an open-source visualization tool called CummeRbund allows for the generation of publication appropriate graphs and figures.

Although this pipeline can be used to assess differentially expressed isoforms, its accuracy and efficiency have been expanded upon by more sophisticated tools. For example, HISAT2 has largely replaced TopHat as the most popular alignment tool in RNA sequencing and Cufflinks has been largely replaced by Stringtie, a next-generation tool designed by many of the same individuals that initially designed the Cufflinks suite. In addition, compatibility and format issues may arise when trying to use outputs from HISAT2 and Stringtie for Cuffdiff 2 analysis. Thus, Cuffdiff 2 is considered to be a deprecated program that is not recommended for differential splicing analysis currently.

4.4.2 DiffSplice

Like Cuffdiff 2, DiffSplice is an isoform-based tool that is used to assess differential alternative splicing patterns across samples. DiffSplice assesses alternative splicing by constructing splice graphs that display the variations in isoforms based on exons that are always either included or excluded. These splice graphs are generated using the reads that were aligned to the reference genome. Two types of read alignments are recognized by the DiffSplice platform and include exonic alignments, which refer to a contiguous sequence of nucleotides that correspond to expressed exonic regions, and spliced alignments, which refer to a contiguous nucleotide sequence that spans at least two exons and has clearly defined acceptor and donor splicing sites. For any given splice graph, represented by the equation $G = (V, E, w)$, every node corresponds to a single exonic unit, which is defined as a genomic region with boundaries delimited by splice sites defined by splice junctions. Because traditional library preparation methods make it difficult to detect the precise location of the start and end transcription sites, these locations are estimated based on where there is a significant change in read coverage from either absent to present or present to absent. Two exonic units will be connected by an edge if there exists a contiguous nucleotide sequence that spans both exons and the directionality of the edge is determined based on the dinucleotide sequence in the intron that flanks the estimated donor and acceptor sites. For instance, a CT–AC dinucleotide pair is indicative of reverse transcription while a GT–AG dinucleotide sequence suggests forward transcription. The DiffSplice tool further optimizes the original splice graph equation $G = (V, E, w)$ by incorporating a transcription start node, denoted ts, and a transcription end node, denoted te. Edges are then added that connect all vertices where transcripts initiate to the ts node as well as connect all vertices where transcripts end to the te node.

Once the augmented splice graph, termed ESG and represented by the formula $G = (V, E, ts, te, w)$, is generated, it then undergoes decomposition to form a series of alternative splicing modules (ASMs). Depiction of the process in which ASMs

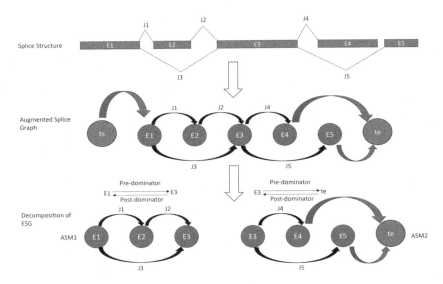

FIGURE 4.7 Overview of the decomposition of the augmented splice graph into multiple ASMs is presented.

are constructed is illustrated in Figure 4.7. An ASM is defined as an induced subgraph of the full ESG and is represented by the following equation (Eq. 4.1):

$$H(ts_H, te_H) = \langle V_H, E_H, ts_H, te_H \rangle \quad (4.1)$$

where ts_H denotes the entry node and te_H denotes the exit node for the ASM. Each separate ASM that is generated must satisfy the following four conditions:

1. **A single entry**: All edges from G to H (all edges within the ASM) originate from ts_H.
2. **A single exit**: All edges from G to H (all edges within the ASM) lead to te_H.
3. **Alternative paths exist**: $d^+(ts_H) > 1$ and $d^-(te_H) > 1$. These are indicative of the number of outgoing edges and incoming edges, respectively.
4. No vertex, denoted uV_H, exists such that the vertex u post-dominates vertex v (meaning that all paths to the *te* node from vertex v contain vertex u) or pre-dominates vertex v (meaning that all paths from the *ts* node to vertex v contain vertex u) in the ESG.

Once the ASMs have been constructed, the DiffSplice algorithm then approximates the number of transcripts that run through each particular splice path for each sample. Ultimately, the algorithm estimates the relative proportion as well as the expression level for each alternative path for each sample. While many algorithms utilize count-based Poisson modeling to collect the number of reads,

Constitutive and Alternative Splicing Events

DiffSplice instead employs a generalized model approach that incorporates the observed support on spice junctions as well as exon expression estimates in order to approximate the abundances for each alternative path. When given a particular transcript, denoted t, and the reads from one sample, c^t_i represents the number of reads that cover the ith nucleotide in transcript t and l_t represents the exonic length of transcript t. The read coverage of transcript t is defined as the average number of reads that cover each nucleotide within the transcript and is represented by the following equation (Eq. 4.2):

$$C_t = \left(\frac{1}{l_t}\right)\sum_{i=1}^{lt} C_i^t \tag{4.2}$$

In this instance, C_t provides a direct estimate of the expression level for transcript t. We can similarly measure the read coverage of an exonic region, e, as well as the read coverage of a splice junction, j, respectively, using the equations below. In these equations, C_e measures the number of transcripts that go through the given exonic sequence e (Eq. 4.3), while C_j provides an estimate of the number of transcript copies that span from the donor exon to the acceptor exon using the junction j (Eq. 4.4).

$$C_e = \left(\frac{1}{l_e}\right)\sum_{i=1}^{le} C_i^e \tag{4.3}$$

$$C_j = \left(\frac{1}{l_j}\right)\sum_{i=1}^{lj} C_i^j \tag{4.4}$$

We can next fit a model that assumes a random sampling process and independence for reads as well as fits a binomial distribution with parameters N_t, the total number of reads from transcript t, and a probability that a read from transcript t falls within an exonic sequence e is represented by the formula $p_{et} = (l_e/l_t)$. When the total number of reads mapped from transcript t is sufficiently large, the number of reads that fall within an exonic sequence e, denoted N_{et}, can be well approximated by the normal distribution with mean $= N_t P_{et}$ and variance $= N_t P_{et}(1 - P_{et})$. A higher degree of coverage and longer segments result in estimators with smaller variances and an overall smaller deviation from the true transcript coverage.

For a given gene G, abundance estimates are calculated beginning with the ASMs located at the bottom of the decomposition hierarchy and then moving upwards to the ASMs located at the top of the hierarchical cluster (Figure 4.8). Statistical testing is performed separately at both the gene level and the transcript level. Differences in expression at the gene level reflect the total expression change of all transcripts derived from a single gene. In contrast, differences observed at the transcript level with regard to the relative proportion of alternative transcript paths reflect the differences in regulatory mechanisms for individual transcripts.

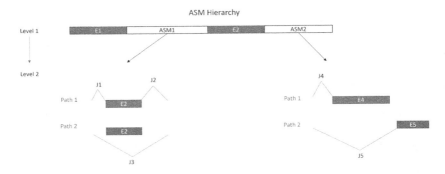

FIGURE 4.8 Overview of the ASM hierarchy that is used to calculate expression estimates by DiffSplice is presented.

Ultimately, statistically significant differences in transcription are determined based on the degree of dissimilarity between the observed and expected test statistics. Therefore, we are most concerned with ASMs with highly divergent path distributions across the two experimental groups while having highly comparable distributions within the same experimental group.

4.4.3 DEXSEQ

Unlike the previously mentioned tools Cuffdiff 2 and DiffSplice which are based on an isoform-based approach, DEXSeq and subsequent differential splicing tools function on an exon-based approach that assesses differential exon usage across samples. The primary data structure for the DEXSeq tool is a table that displays the number of reads that overlap with a particular exon. This structure is repeated for all exons for every gene from our samples. However, for exons that have differing boundaries across transcript species, i.e., alternative splice sites, the exon is cut into two or more pieces and is placed into what is termed a counting bin. It is important to note that any read that overlaps with multiple counting bins for a given gene is counted in each bin. Figure 4.9 illustrates this so-called flattening of gene models approach employed by the DEXSeq tool. In this figure, each orange rectangle represents a single exon while each green rectangle represents a single counting bin. For exons that are uniform in length across all transcripts, a single counting bin is created. In contrast, exons with variable lengths due to alternative splicing events are divided into two counting bins.

Let k_{ijl} denote the number of reads that overlap a given counting bin l for a particular gene i in sample j and let μ_{ijl} represent the expected count of cDNA fragments that are a part of counting bin l for gene i. The expected read count, denoted $E(k_{ijl})$, relates to μ_{ijl} via the size factor S_j, a function that takes into account the sequencing depth of a given sample j. The overall relationship between the expected cDNA fragment count and the expected read count is explained via the formula $E(k_{ijl}) = S_j\mu_{ijl}$. Generalized linear models are fitted in order to model

Constitutive and Alternative Splicing Events

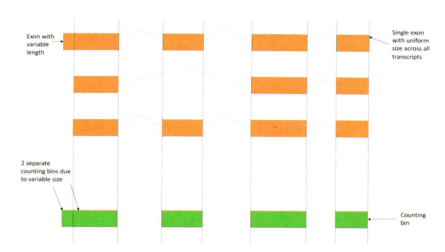

FIGURE 4.9 Flattening of the gene models is presented above. Each set of dotted lines denotes a single counting bin. Exons that have different boundaries across transcripts are divided into multiple counting bins.

read counts and we assume that the number of overlapping reads k_{ijl} adheres to a negative binomial distribution with *mean* = $S_j \mu_{ijl}$ and *dispersion* = α_{il}. The mean for each bin can subsequently be predicted using the following log-linear model (Eq. 4.5):

$$\log \mu_{ijl} = \beta_i^G + \beta_{il}^E + \beta_{ipi}^C + \beta_{ipil}^{EC} \qquad (4.5)$$

where β_i^G denotes the baseline expression level for a given gene i, β_{il}^E represents the log of the expected read fragments that map to gene i and overlap with bin l, β_{ipi}^C denotes the log fold change in expression for a given gene i for experimental condition (sample) p_i, and β_{ipil}^{EC} represents the effect that a given experimental condition p_i has on influencing the number of reads that fall into counting bin l. The terms of particular interest in this model are β_{ipi}^C and β_{ipil}^{EC}. A value for β_{ipi}^C that is significantly different from 0 suggests the counting bin that it is associated with is differentially utilized while a value for β_{ipil}^{EC} that significantly differs from 0 is indicative of an overall difference in expression for a particular gene. Model parameters are calculated using the modified approach iteratively reweighted least square algorithm and dispersion estimates are calculated using the Cox–Reid correction. Ultimately, we test against the null hypothesis that states there is no significant difference in the number of reads that overlap a particular counting bin l between conditions. Thus, for each gene, we fit a reduced model missing the interaction term (β_{ipil}^{EC}) and compare it to a full model containing an interaction term for each counting bin l for gene i. Because differential exon usage cannot be parsed from overall differential expression (not taking into account differential exon usage) if a given gene only has 1 counting bin or if a given gene has multiple

114 RNA-seq in Drug Discovery and Development

counting bins and all but 1 have 0 counts, all counting bins with 0 counts across all samples as well as all counting bins for genes with less than 2 nonzero bins are denoted as *not testable* by the DEXSeq algorithm.

In order to illustrate the output and overall analysis performed by DEXSeq, we will utilize a public dataset that deep sequenced S2-DRSC cells derived from the model organism *Drosophila melanogaster* (common fruit fly) that were RNAi depleted of mRNAs that encode for various RNA binding proteins, particularly the CG7878 gene in the case of this example (GEO accession: GSE18508). Raw reads (.fasta/.fastq) single-read files, one of which was a control sample and two of which corresponded to cells treated with dsRNA to knockdown CG7878-derived mRNAs, were downloaded from the NCBI gene expression omnibus (GEO) database and then uploaded into the Galaxy interface. Raw reads were then aligned to the reference genome (dm3) using the HISAT2 alignment tool in order to generate our .bam files using the parameters presented in Figure 4.10. Unlike previous examples that used the alignment tool HISAT2, we will modify the *Splice Alignment options* option underneath the Advanced Options tab. We will leave all of these options at their default values except for the *GTF file with known splice sites*, which will use a separate Ensembl-derived annotated reference genome for *Drosophila*. For consistency, the following example will use a form of the annotated reference genome that is pre-processed by DEXSeq; however, the standard. gtf downloaded from http://ftp.ensembl.org/pub/release-105/gtf/drosophila_melanogaster/ can be used in its stead with no complications.

Once we have our alignment files and our reference.gtf file, we can now utilize the DEXSeq tool. First, we must convert our standard annotated reference.gtf file into what is termed a 'flattened GTF' file, which contains discrete, well-defined, non-overlapping exon bin definitions with annotations for each transcript. To accomplish this conversion, we will utilize the DEXSeq-Count tool in Galaxy. The DEXSeq-Count tool has two modes of operations, the first of which is termed Prepare annotation and is used to convert .gtf files into ones that are compatible with DEXSeq while the second mode is termed Count Reads which generates raw count tables and estimates exon abundances. We will first use the Prepare annotation mode using our downloaded Ensembl reference genome with the parameters defined in Figure 4.11.

Once the flattened GTF file has been produced, we can then switch from the Prepare annotation mode to the Count reads mode and generate our DEX-Seq compatible count matrices using our bam files produced from HISAT2 and our recently created flattened GTF file (Figure 4.12). It is important to check the count matrices that are generated in order to ensure that the data makes logical sense. Due to either compatibility issues or differences in gene labels, it is possible for the job to run to completion yet all of the counts in the matrices are equal to 0. In this case, using the count matrices in the DEXSeq tool will generate an error message relating to all genes having a value of 0 in their counting bins. For this reason, it is recommended that you use the same reference genome, preferably from Ensembl, at every step of the analysis to ensure proper labeling. More complex methods are available to modify these files to make them formatted in a way that is compatible with DEXSeq; however, these methods are beyond the scope of this book and will not be discussed in detail.

Constitutive and Alternative Splicing Events

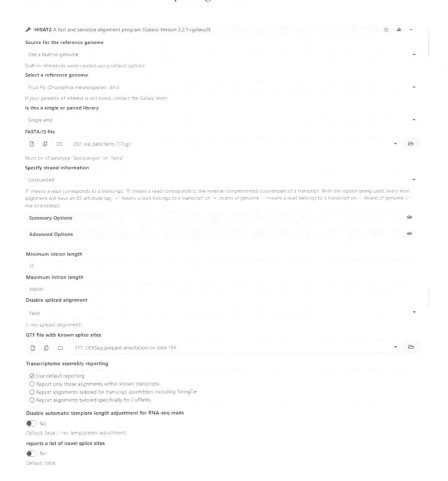

FIGURE 4.10 HISAT2 parameters used to generate alignment files for the *Drosophila* alternative splicing analysis.

FIGURE 4.11 Parameters used for the Prepare annotation mode of the DEXSeq-Count tool in order to generate our flattened GTF file.

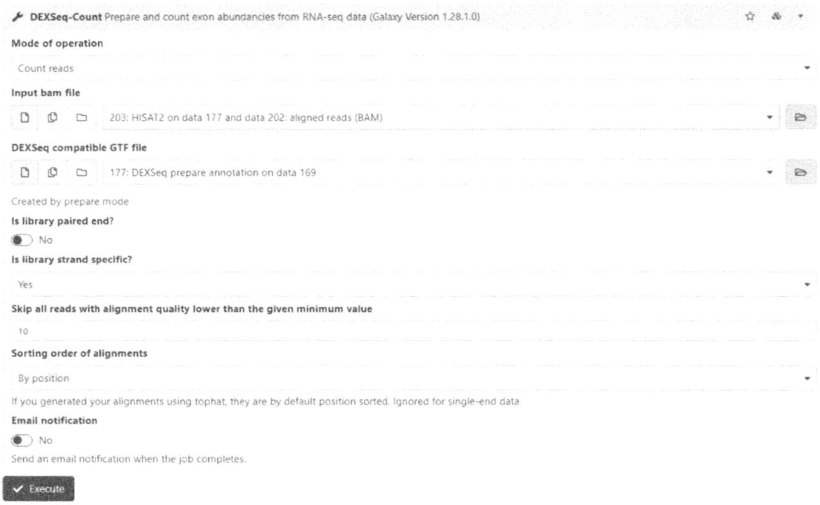

RNA-seq in Drug Discovery and Development

FIGURE 4.12 Parameters used for the Count reads mode of the DEXSeq-Count tool in order to generate count matrices that are compatible with the DEXSeq tool.

Once the count matrices have been generated for each alignment file, we can then utilize the DEXSeq tool to analyze differential exon usage for all genes in our sample. As we performed in other differential analysis tools, we will denote our factor of distinction as 'Treatment status' and the two levels of this factor as 'Control' and 'Treated', respectively. As input, we will utilize our flattened GTF file that was prepared using the Prepare annotation mode of the DEXSeq-Count tool and our .bam files separated into their respective factor level. An additional modification that we will make to the default settings of the tool is changing the Output rds file for plotDEXSeq from *no* to *yes* (Figure 4.13). This will generate a separate RData file that can be used to visualize the results from the DEXSeq tool using the plotDEXSeq tool in Galaxy. Submitting the DEXSeq job will result in the following three outputs:

1. A tabular result file containing a list of DEGs and differentially utilized exons across samples including fold change and a list of transcripts that overlap with the particular exon (Table 4.1).
2. An html report that can be downloaded directly to your computer and contains graphical displays for each DEG with exons that are utilized differentially (alternative splicing graphs).
3. An rds file in RData format that can be used with the plotDEXSeq tool for enhanced visualization of alternative splicing.

The html file that can be downloaded includes four distinct graphs for each gene with at least one instance of differential exon usage. The four graphs include a

Constitutive and Alternative Splicing Events

FIGURE 4.13 Parameters used in the DEXSeq tool to generate output analysis of differential exon usage.

counts.svg file, which shows normalized counts for each exon bin in each sample, an expression.svg file, which shows the average expression of counts across all samples for each exon bin, a splicing.svg file, which shows the average exon usage across all samples, and a transcript.svg file, which shows differential expression of each transcript that mapped to the gene of interest (each isoform). In each file, a graph displaying the expression/exon usage (Y-axis) for each exon comprising the selected gene (X-axis) is presented for each factor level in the experiment. In addition, a flattened representation of the gene including splice junctions and all exons is presented below each graph showing the respective counts/expression of each exon bin. In each graph type, a purple exon in the flattened model denotes an exon that is differentially used across factor levels. Figures 4.14 and 4.15 display sample graphs for the genes *FBgn0000042* and *FBgn0000173*, respectively.

TABLE 4.1
Overall Layout of the Tabular Results File Generated by the DEXSeq Tool

Column	Description
1	Identifiers for genes and exons
2	Group/gene identifier
3	Feature/exon identifier
4	Mean counts across samples for each feature
5	Exon dispersion estimate
6	Likelihood-ratio test statistic
7	Likelihood-ratio test *p*-value
8	Adjusted *p*-values
9	Exon usage coefficient for factor level 2
10	Exon usage coefficient for factor level 1
11	Relative exon usage fold changes
12	Coordinates of the feature
13	Matrix of integer counts
14	List of transcripts that overlap with the exon

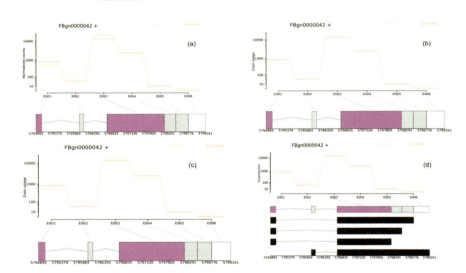

FIGURE 4.14 The four graphs from the html output from DEXSeq for the gene *FBgn0000042* including for counts (a), expression (b), splicing (c), and transcripts (d) are presented.

4.4.4 EDGER

DEXSeq is a downstream analysis tool that can be used to assess differential exon usage. This is distinct from tools such as edgeR and *limma* that measure absolute differences in overall gene expression as DEXSeq identifies differential splicing

Constitutive and Alternative Splicing Events

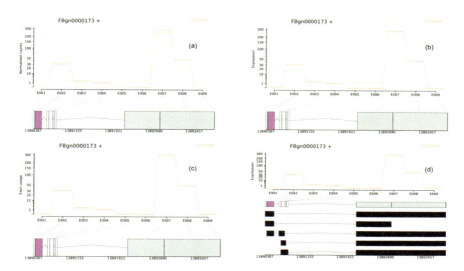

FIGURE 4.15 The four graphs from the html output from DEXSeq for the gene *FBgn0000173* including for counts (a), expression (b), splicing (c), and transcripts (d) are presented.

events that lead to changes in the relative abundances of each possible transcript for a given gene. In contrast, *limma* and edgeR compare the absolute expression of an isoform in one condition to the absolute expression of the same isoform in the second condition. Suppose we have a classic two-condition experiment with 10,000 transcripts generated for a given gene in the control experimental group and 15,000 transcripts generated for that same gene in the experimental group. Based on this set of results, edgeR and *limma* would detect differential expression for this particular gene between the two conditions.

Now let us look at the data from a different perspective. Let the absolute number of transcripts for a given gene be the same as previously stated (10,000 transcripts for the control group and 15,000 transcripts in the experimental group) with five possible isoforms for the gene that have the following relative abundances in each group: isoform 1 (20%), isoform 2 (20%), isoform 3 (20%), isoform 4 (20%), and isoform 5 (20%). In this case, the absolute number of each isoform (2,000 vs. 3,000 when comparing the control group with the treated group) changes yet the relative abundances of each isoform do not change across samples. As a result, DEXSeq would not detect a significant difference in exon usage for this gene across samples despite the fact that absolute differences in gene expression were observed for this particular gene.

If we recall from Chapter 3, edgeR takes a raw count matrix with each row corresponding to a particular gene and each column referring to a particular sample. Each entry in the matrix represents a count for a given gene in a given sample and is denoted by k_{ij}. The edgeR algorithm assumes an overdispersed Poisson distribution for read counts and the variance is defined as a linear function of the mean.

120 RNA-seq in Drug Discovery and Development

To illustrate, we will utilize a public dataset assessing the efficacy of optimized drug combinations (regorafenib, erlotinib, and vemurafenib) on six various colorectal cancer cell lines relative to the traditional FOLFOX regimen, which consists of the drugs folinic acid, fluorouracil, and oxaliplatin (GEO accession: GSE142340). For our purposes, we will assess alternative splicing in the human colorectal adenocarcinoma DLD-1 cell line by comparing control cells, which were treated with the standard FOLFOX chemotherapy treatment, and cells treated with the aforementioned optimized drug combination of regorafenib, erlotinib, and vemurafenib. Raw reads (.fasta/.fastq) files were downloaded from the NCBI GEO database and then uploaded into the Galaxy interface. Raw reads were then aligned to the reference genome (hg38) using the HISAT2 alignment tool (Figure 4.16a) in order to generate our .bam files and raw count matrices were generated using *featurecounts* with the parameters presented in Figure 4.16b. All .fastq files were processed using the presented parameters in Figure 4.16a and b for HISAT2 and *featurecounts,* respectively. The factor of interest was designated as treatment status while the contrast of interest was denoted as control-treated and each separate count file was then grouped, respectively, depending on whether or not the sample was treated with FOLFOX or the optimized drug combination (Figure 4.17).

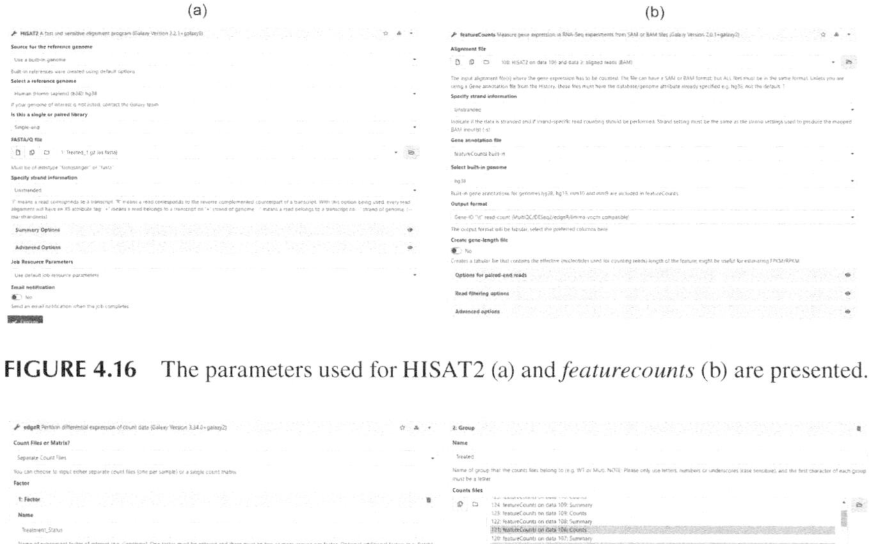

FIGURE 4.16 The parameters used for HISAT2 (a) and *featurecounts* (b) are presented.

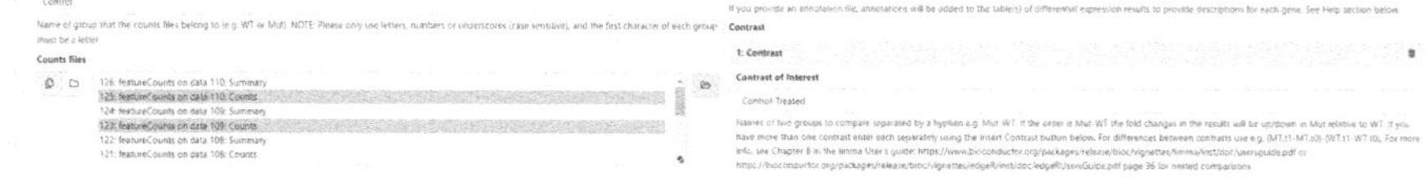

FIGURE 4.17 The parameters used for the control-treated contrast in edgeR are presented.

Using an FDR cutoff of significance 5%, 326 unique gene ids were found to be differentially expressed between the control samples and the treated samples, an excerpt of which is shown in Figure 4.18.

Figure 4.19 presents the output report files generated after completion of the edgeR job including the biological coefficient of variance (BCV) plot, quasi-likelihood (QL) plot, and mean-difference (MD) plot. When the list of significant DEGs was inputted into the database for annotation, visualization, and integrated discovery (DAVID) tool, genes were mapped to ten distinct molecular Kyoto encyclopedia of genes and genomes (KEGG) mechanisms with the top five including MAPK signaling, p53 signaling, cellular senescence, transcriptional misregulation in cancer, and bladder cancer (Figure 4.20). If we take a closer look at the most

GeneID	logFC	logCPM	F	PValue	FDR
1192	12.2604279419869	3.91081286652009	942.95395309901	5.35078127655686e-14	1.51935434347832e-09
84951	2.2873148486645	7.36427463744933	547.791578455036	5.10961926610111e-13	7.25438195304705e-09
8061	2.36324175404122	5.8111155403073	485.406192834622	1.21111290349897e-12	1.14631836316178e-08
1847	2.33605891466107	4.85796088030771	304.921817550835	3.24631453045487e-11	2.30447752730665e-07
1844	2.41325050452137	3.68293990249737	217.364690401056	3.41854488555978e-10	1.9413916405094e-06
1649	-1.84917402753582	4.35742139505983	197.113244549377	6.69467510558023e-10	3.16825499371584e-06
81848	3.99149168394463	2.58185840291933	191.73154019951	8.09135973713409e-10	3.28220228194175e-06
2152	2.17358004129413	6.20100288218532	308.103409106502	1.39502493863934e-09	4.95146664158301e-06
4144	1.12656017481547	7.80911969244232	165.302810762401	2.21952185935592e-09	6.70720607705075e-06
3726	1.37309118833794	5.25257229075597	163.788275602677	2.36210814476167e-09	6.70720607705075e-06
9075	1.01713076208642	6.58932871982064	138.868989508293	7.15758295984088e-09	1.84763243767893e-05
6004	1.98855812690514	3.23006598269276	120.308430037319	1.85185819469314e-08	4.38195945319264e-05
79170	-1.36622309922529	4.89669974114208	113.265656761353	2.75068714300587e-08	6.00813549428089e-05
4665	1.212667188371	5.29905650440744	111.413209090768	3.06354370190213e-08	6.21352310110792e-05
54407	-0.837568489803932	7.75938480907205	108.919415598345	3.55067532710773e-08	6.72142839421494e-05
2827	2.43665385444311	2.49040585630802	106.262100397395	4.16919475496565e-08	7.39901781670311e-05
9709	-0.923507345636047	7.06879726599218	103.409448086967	4.97350655185669e-08	8.30721873764569e-05
10769	0.906556974539132	6.19535092926602	100.822336668522	5.85834646282652e-08	9.24154154510883e-05
23143	1.02167617652529	5.62990689696473	98.2996236162523	6.89736937381519e-08	0.000102325531186837
6385	0.805058650582532	7.2890078713238	97.6300632031587	7.20729226049443e-08	0.000102325531186837
25805	-0.99606563199443	5.35821080748203	92.734657235133	1.00214791304643e-07	0.000135504714242635
9020	0.835856432861298	6.25003126719194	90.1205753449633	1.20254824093066e-07	0.000145332266484058
161742	1.04550312903019	5.03887762935006	90.1161103360817	1.20292745405037e-07	0.000145332266484058
23764	2.56284478343681	3.89047354364286	147.294030467048	1.22837626188321e-07	0.000145332266484058
6376	1.5480397892487	3.4247540119483	87.6712620970466	1.43257133620631e-07	0.000155320239165201
1831	-2.26520272783484	4.65457112802433	158.878171209759	1.46036504657535e-07	0.000155320239165201
3815	-0.759390848245844	6.96615035814854	87.2508013377037	1.47689609348844e-07	0.000155320239165201
9816	0.833537214284845	6.22310887373411	84.7595931558458	1.77382130293699e-07	0.000179884485346057
200734	0.804653349553456	6.68954458667076	79.7993484564552	2.59126240289658e-07	0.000253720330079396
2069	0.801895464639685	9.31937729839233	78.2433120832034	2.93058862852765e-07	0.000277380213690142
6574	1.24775064660142	7.10799135292629	123.632168790075	3.15150591955596e-07	0.000283117342773333
27242	0.898723003515104	7.09265642593381	76.753072913749	3.30365912958278e-07	0.000283117342773333
26579	1.22497307305645	4.2017605172117	76.7193801598953	3.31269703933593e-07	0.000283117342773333
5723	-0.78556955032106	6.64727935506352	76.4352923826908	3.39002981309855e-07	0.000283117342773333
100507056	1.32729103391506	4.17230228646632	75.3809243991041	3.69553816339944e-07	0.000299813569598812
56603	1.4952207803755	3.2459289380028	74.9830173284846	3.81887464604395e-07	0.000301213737264665
6461	1.03239687184853	5.45056698863748	72.968911848663	4.15494450207722e-07	0.000318863916584703
57050	0.888904200764529	5.78149702050578	72.4829534775819	4.70991597707801e-07	0.000340833998883788
8797	0.931284784193414	5.17991791897466	72.4062151216328	4.74078742011521e-07	0.000340833998883788
729438	-1.03511805627662	4.8049351038219	72.2216983307721	4.81596356735566e-07	0.000340833998883788
60485	0.866589617729297	5.75412758471429	71.8793387030091	4.95906657450062e-07	0.000340833998883788
468	-0.721585038659121	9.32719361902213	71.6874374604299	5.04139036844196e-07	0.000340833998883788
56652	0.807189593964405	5.65683797925608	70.0794558369755	5.79570238812467e-07	0.000382718533328093
2526	0.767734240787095	5.88843804110602	69.770115200092	5.95510305745588e-07	0.000384307162082863
7975	1.1674140243688	4.19631584591697	68.8504897917509	6.45921107670765e-07	0.000407576218940253
7422	-0.791531590395627	6.1179883718574	68.540327937069	6.64004129466222e-07	0.000409878201221595
993	0.741704511147973	6.71887569956113	66.9590399648002	7.64935266130749e-07	0.000462134827271971
24147	1.44042768633586	3.58215866839099	66.302595359769	8.12930118690308e-07	0.000480898973337735
9052	0.748040808551041	7.73398131017508	64.9255132639605	9.23357241159118e-07	0.000535076099239044
54931	0.935496629188802	5.13413034228294	63.6181501433256	1.04422203527193e-06	0.000591254561039978
390	0.916988787530285	5.47678961153267	63.4409668903572	1.06194691364814e-06	0.000591254561039978
1848	3.45169666242368	6.76410715274186	368.75419975397	1.09589633523654e-06	0.000598422623827721
30836	0.658129313848368	7.05002432229275	62.432395657545	1.16959789622147e-06	0.000626617589871859
3475	-0.765359711142345	5.94249841333773	62.2020933610782	1.19588698280867e-06	0.000628837238460228
84879	0.944359200365848	4.63867648277442	61.946674925165	1.22583195381346e-06	0.000632863605973332
23268	0.774259170801248	5.89073342427598	61.677938457622	1.25826222576283e-06	0.000633542192858326
6996	0.784395746407374	5.85749827379648	61.5683200891288	1.27176985359833e-06	0.000633542192858326
201164	1.18233262156307	3.64012326314076	60.8247988893353	1.36786439948526e-06	0.000669663959023862
3344	0.895445936450475	4.87704896041072	60.5051781238483	1.41169027850514e-06	0.000679405855222294
10797	-0.682486397477144	8.25443666775196	59.777309422424	1.51758271944775e-06	0.000718196021978648
1960	2.83948374738675	0.968289525314141	59.0818879213797	1.62726488552876e-06	0.000757478465976873
81875	0.669358296004657	6.22281392599572	58.6262957602249	1.7040041567734e-06	0.000776129125814125
84706	-0.761400633247421	6.12103651980713	58.5228409046264	1.72199806044338e-06	0.000776129125814125

FIGURE 4.18 Sample excerpt from the list of significant DEGs from the edgeR output.

Differential Expression Counts:

	Up	Flat	Down
Control-Treated	252	28069	74

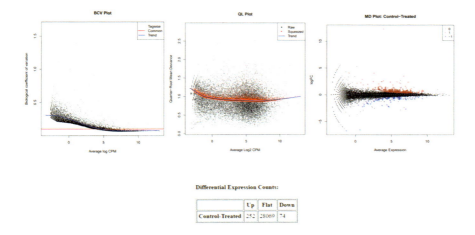

FIGURE 4.19 Output report generated from edgeR including BCV, QL, and MD plots.

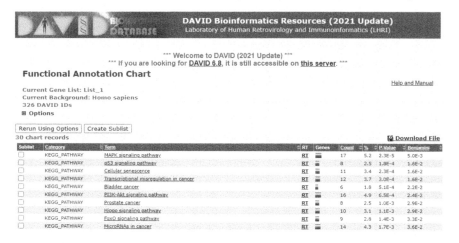

FIGURE 4.20 List of significant molecular KEGG pathways identified by the DAVID tool using functional enrichment analysis.

statistically significant MAKP signaling pathway, we find that 17 genes that map to this pathway are differentially expressed. The MAPK signaling pathway and the list of DEGs that map to this pathway are presented in Figure 4.21.

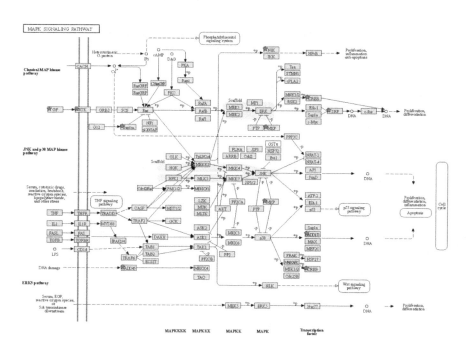

FIGURE 4.21 KEGG pathway for the MAPK signaling pathway.

Constitutive and Alternative Splicing Events

4.4.5 LIMMA

Like edgeR, linear models for microarray data (*limma*), particularly *limma-voom* for RNA sequencing data, assesses differential expression of genes across different treatment conditions. Unlike edgeR however, *limma-voom* models raw expression counts using linear models in order to estimate intrasample and intersample variability. If we recall from Chapter 3, *limma*, like edgeR, requires a matrix of raw read counts as input with columns denoting unique samples and rows representing genomic features. A linear model is fitted for each row and regression coefficients, standard errors, test statistics, and *p*-values are estimated for each comparison of interest.

The following example will utilize samples from the same public dataset used for the DEXSeq that deep sequenced S2-DRSC cells derived from the model organism *Drosophila melanogaster* (common fruit fly). However, we will use three replicates for both the control and treated groups, two of which are paired-end and one of which is single-end. In addition, we will utilize cell samples that were RNAi depleted of mRNAs that encode for the CG8144 gene (GEO accession: GSE18508). Raw reads (.fasta/.fastq) were downloaded from the NCBI GEO database and then uploaded into the Galaxy interface. Raw reads were then aligned to the reference genome (dm3) using the HISAT2 alignment tool in order to generate our .bam files using the same parameters presented in Figure 4.10. Like the DEXSeq example, we modified the *Splice Alignment options* option underneath the Advanced Options tab and left all of these options at their default values except for the *GTF file with known splice sites*, which we again used our Ensembl-derived annotated reference genome for *Drosophila*. Raw count matrices were generated using *featurecounts* and the parameters are presented in Figure 4.22. The raw count matrices generated from *featurecounts* were subsequently used as input for *limma* using the parameters presented in Figure 4.23, and the output graphs from the *limma* report analysis are presented in Figure 4.24.

At a significance level of 0.05 (Adj-*p*-value ≤ 0.05), 907 genes in total were found to be differentially expressed between the control and treated groups and mapped to four distinct KEGG pathways including extracellular matrix (ECM)–receptor interaction (7 genes), biosynthesis of amino acids (15 genes), metabolic pathways (106 genes), and carbon metabolism (19 genes) (Figure 4.25). Figure 4.26 shows the ECM–receptor interaction pathway as well as the list of seven genes that are differentially expressed across samples.

4.5 SUMMARY

In this chapter, we introduced the concept of alternative splicing mechanisms as well as tools that can be used to identify and assess alternative splicing across samples. Tools such as cuffdiff2 and DiffSplice are isoform-based and can be used to make inferences concerning differences in alternative splicing events across samples. The second major approach for differential splicing analysis was also discussed and included exon-based approaches including DEXSeq,

FIGURE 4.22 *Featurecounts* parameters used to generate count matrices for each alignment file for the RNAi *Drosophila* experiment.

limma, and edgeR. DEXSeq can be used to assess relative abundances of transcripts across factor levels to infer differential exon usage. In contrast, tools such as *limma* and edgeR measure the absolute changes in gene expression for a single transcript across conditions and then repeat for every mapped read from the experiment.

KEYWORDS AND PHRASES

After reading this chapter, you should be able to demonstrate familiarity with the following words and phrases:

- Understand what alternative splicing is and the mechanism behind the phenomenon.
- Understand how the Cuffdiff 2 algorithm can be used to identify alternative splicing events.
- Be able to use Galaxy to replicate the differential expression and alternative splicing analyses with Cuffdiff, DiffSplice, DEXSeq, edgeR, and Limma.
- Understand how isoform-based approaches (Cuffdiff and DiffSplice) differ from exon-based approaches (DEXSeq, edgeR, and Limma).

Constitutive and Alternative Splicing Events 125

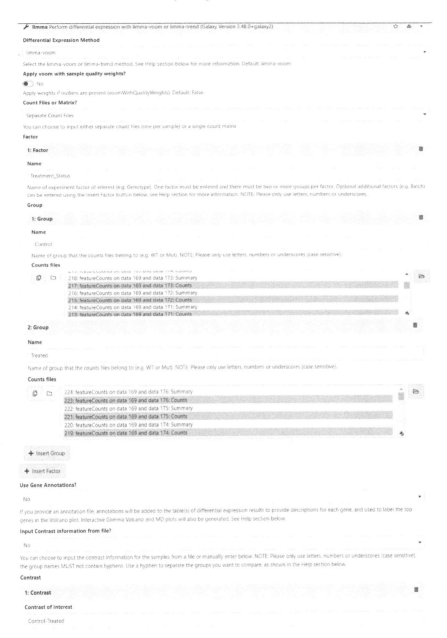

FIGURE 4.23 LIMMA parameters used for differential expression analysis in the RNAi *Drosophila* experiment.

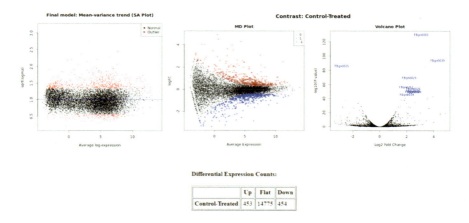

FIGURE 4.24 LIMMA output including the mean-variance sensitivity analysis (SA) plot, mean-difference (MD) plot of average expression, volcano plot, and the number of differentially expressed genes are presented.

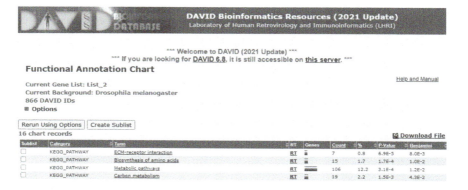

FIGURE 4.25 Significantly enriched KEGG pathways identified by the DAVID bioinformatics tool.

BIBLIOGRAPHY

Anders, S., Reyes, A., & Huber, W. (2012, Oct). Detecting differential usage of exons from RNA-seq data. *Genome Res, 22*(10), 2008–2017. https://doi.org/10.1101/gr.133744.111

Brooks, A. N., Aspden, J. L., Podgornaia, A. I., Rio, D. C., & Brenner, S. E. (2011, Oct). Identification and experimental validation of splicing regulatory elements in Drosophila melanogaster reveals functionally conserved splicing enhancers in metazoans. *RNA, 17*(10), 1884–1894. https://doi.org/10.1261/rna.2696311

Corvelo, A., Hallegger, M., Smith, C. W., & Eyras, E. (2010, Nov 24). Genome-wide association between branch point properties and alternative splicing. *PLoS Comput Biol, 6*(11), e1001016. https://doi.org/10.1371/journal.pcbi.1001016

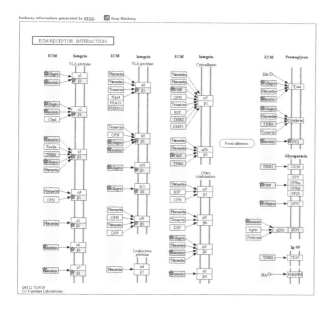

FIGURE 4.26 ECM–receptor interaction KEGG pathway and the list of DEGs that mapped to the pathway.

Ding, F., & Elowitz, M. B. (2019, Jun). Constitutive splicing and economies of scale in gene expression. *Nat Struct Mol Biol, 26*(6), 424–432. https://doi.org/10.1038/s41594-019-0226-x

Dvinge, H. (2018, Sep). Regulation of alternative mRNA splicing: Old players and new perspectives. *FEBS Lett, 592*(17), 2987–3006. https://doi.org/10.1002/1873-3468.13119

Hu, Y., Huang, Y., Du, Y., Orellana, C. F., Singh, D., Johnson, A. R., Monroy, A., Kuan, P. F., Hammond, S. M., Makowski, L., Randell, S. H., Chiang, D. Y., Hayes, D. N., Jones, C., Liu, Y., Prins, J. F., & Liu, J. (2013, Jan). DiffSplice: The genome-wide detection of differential splicing events with RNA-seq. *Nucleic Acids Res, 41*(2), e39. https://doi.org/10.1093/nar/gks1026

Huang da, W., Sherman, B. T., & Lempicki, R. A. (2009a, Jan). Bioinformatics enrichment tools: paths toward the comprehensive functional analysis of large gene lists. *Nucleic Acids Res, 37*(1), 1–13. https://doi.org/10.1093/nar/gkn923

Huang da, W., Sherman, B. T., & Lempicki, R. A. (2009b). Systematic and integrative analysis of large gene lists using DAVID bioinformatics resources. *Nat Protoc, 4*(1), 44–57. https://doi.org/10.1038/nprot.2008.211

Kanehisa, M. (2019, Nov). Toward understanding the origin and evolution of cellular organisms. *Protein Sci, 28*(11), 1947–1951. https://doi.org/10.1002/pro.3715

Kanehisa, M., Furumichi, M., Sato, Y., Ishiguro-Watanabe, M., & Tanabe, M. (2021, Jan 8). KEGG: Integrating viruses and cellular organisms. *Nucleic Acids Res, 49*(D1), D545–D551. https://doi.org/10.1093/nar/gkaa970

Kanehisa, M., & Goto, S. (2000, Jan 1). KEGG: Kyoto encyclopedia of genes and genomes. *Nucleic Acids Res, 28*(1), 27–30. https://doi.org/10.1093/nar/28.1.27

Law, C. W., Chen, Y., Shi, W., & Smyth, G. K. (2014, Feb 3). Voom: Precision weights unlock linear model analysis tools for RNA-seq read counts. *Genome Biol, 15*(2), R29. https://doi.org/10.1186/gb-2014-15-2-r29

Matera, A. G., & Wang, Z. (2014, Feb). A day in the life of the spliceosome. *Nat Rev Mol Cell Biol, 15*(2), 108–121. https://doi.org/10.1038/nrm3742

McCarthy, D. J., Chen, Y., & Smyth, G. K. (2012, May). Differential expression analysis of multifactor RNA-Seq experiments with respect to biological variation. *Nucleic Acids Res, 40*(10), 4288–4297. https://doi.org/10.1093/nar/gks042

Mehmood, A., Laiho, A., Venalainen, M. S., McGlinchey, A. J., Wang, N., & Elo, L. L. (2020, Dec 1). Systematic evaluation of differential splicing tools for RNA-seq studies. *Brief Bioinform, 21*(6), 2052–2065. https://doi.org/10.1093/bib/bbz126

Ritchie, M. E., Diyagama, D., Neilson, J., van Laar, R., Dobrovic, A., Holloway, A., & Smyth, G. K. (2006, May 19). Empirical array quality weights in the analysis of microarray data. *BMC Bioinformatics, 7*, 261. https://doi.org/10.1186/1471-2105-7-261

Ritchie, M. E., Phipson, B., Wu, D., Hu, Y., Law, C. W., Shi, W., & Smyth, G. K. (2015, Apr 20). limma powers differential expression analyses for RNA-sequencing and microarray studies. *Nucleic Acids Res, 43*(7), e47. https://doi.org/10.1093/nar/gkv007

Robinson, M. D., McCarthy, D. J., & Smyth, G. K. (2010, Jan 1). EdgeR: A Bioconductor package for differential expression analysis of digital gene expression data. *Bioinformatics, 26*(1), 139–140. https://doi.org/10.1093/bioinformatics/btp616

Robinson, M. D., & Smyth, G. K. (2007, Nov 1). Moderated statistical tests for assessing differences in tag abundance. *Bioinformatics, 23*(21), 2881–2887. https://doi.org/10.1093/bioinformatics/btm453

Robinson, M. D., & Smyth, G. K. (2008, Apr). Small-sample estimation of negative binomial dispersion, with applications to SAGE data. *Biostatistics, 9*(2), 321–332. https://doi.org/10.1093/biostatistics/kxm030

Trapnell, C., Hendrickson, D. G., Sauvageau, M., Goff, L., Rinn, J. L., & Pachter, L. (2013, Jan). Differential analysis of gene regulation at transcript resolution with RNA-seq. *Nat Biotechnol, 31*(1), 46–53. https://doi.org/10.1038/nbt.2450

Trapnell, C., Roberts, A., Goff, L., Pertea, G., Kim, D., Kelley, D. R., Pimentel, H., Salzberg, S. L., Rinn, J. L., & Pachter, L. (2012, Mar 1). Differential gene and transcript expression analysis of RNA-seq experiments with TopHat and Cufflinks. *Nat Protoc, 7*(3), 562–578. https://doi.org/10.1038/nprot.2012.016

Trapnell, C., Williams, B. A., Pertea, G., Mortazavi, A., Kwan, G., van Baren, M. J., Salzberg, S. L., Wold, B. J., & Pachter, L. (2010, May). Transcript assembly and quantification by RNA-Seq reveals unannotated transcripts and isoform switching during cell differentiation. *Nat Biotechnol, 28*(5), 511–515. https://doi.org/10.1038/nbt.1621

Wagner, B., Riggs, P., & Mikulich-Gilbertson, S. (2015). The importance of distribution-choice in modeling substance use data: A comparison of negative binomial, beta binomial, and zero-inflated distributions. *Am J Drug Alcohol Abuse, 41*(6), 489–497. https://doi.org/10.3109/00952990.2015.1056447

Wang, Y., Liu, J., Huang, B. O., Xu, Y. M., Li, J., Huang, L. F., Lin, J., Zhang, J., Min, Q. H., Yang, W. M., & Wang, X. Z. (2015, Mar). Mechanism of alternative splicing and its regulation. *Biomed Rep, 3*(2), 152–158. https://doi.org/10.3892/br.2014.407

Zhang, Y., Qian, J., Gu, C., & Yang, Y. (2021, Feb 24). Alternative splicing and cancer: A systematic review. *Signal Transduct Target Ther, 6*(1), 78. https://doi.org/10.1038/s41392-021-00486-7

Zhou, X., Wu, W., Li, H., Cheng, Y., Wei, N., Zong, J., Feng, X., Xie, Z., Chen, D., Manley, J. L., Wang, H., & Feng, Y. (2014, Apr). Transcriptome analysis of alternative splicing events regulated by SRSF10 reveals position-dependent splicing modulation. *Nucleic Acids Res, 42*(6), 4019–4030. https://doi.org/10.1093/nar/gkt1387

5 The Role of Transcriptomics in Identifying Fusion Genes and Chimeric RNAs in Cancer

Robert Morris, Valeria Zuluaga, and Feng Cheng
University of South Florida

CONTENTS

5.1 Fusion Gene 130
    5.1.1 What Is a Fusion Gene 130
    5.1.2 Mechanisms That Generate New Fusion Genes 130
    5.1.3 Fusion RNA Transcripts 131
    5.1.4 The Connection between Fusion Genes and
           Non-Coding RNAs 133
5.2 Detection Methods for Identification of Fusion Genes and Chimeric
    Proteins 135
    5.2.1 Guided Detection Approaches 135
    5.2.2 High-Throughput Sequencing-Based Detection Methods 136
    5.2.3 ChimPipe 137
           5.2.3.1 Exhaustive Paired-End and SPLIT Read Mapping
                    with Genome Multitool (GEM) 137
           5.2.3.2 ChimSplice 138
           5.2.3.3 ChimPE 138
           5.2.3.4 ChimFilter 138
    5.2.4 GFusion 139
           5.2.4.1 The Beginning Alignment 139
           5.2.4.2 Anchors 140
           5.2.4.3 Alignment and Localization 141
           5.2.4.4 Fusion Boundaries 141
           5.2.4.5 Confirming Fusion Models 141
           5.2.4.6 Grouping Candidate Fusions 141
           5.2.4.7 Fusion Index and Realignment 142

DOI: 10.1201/9781003174028-5

5.2.5	InFusion	142
	5.2.5.1 Short Reads Alignment	142
	5.2.5.2 Local Short Reads Alignment	143
	5.2.5.3 Local Alignment Analysis	144
	5.2.5.4 Analysis of End-to-End Alignments	144
	5.2.5.5 Clustering and Establishing Putative Fusions	144
	5.2.5.6 Purifying and Filtering Fusions	145
	5.2.5.7 Reporting Fusions	145
5.2.6	STAR-Fusion	145
5.3	Summary	149
	Keywords and Phrases	149
	Bibliography	149

5.1 FUSION GENE

5.1.1 WHAT IS A FUSION GENE

The human genome consists of more than 3 billion base pairs, which are the building blocks of the approximately 25,000 genes comprising the human genome. A gene on its own is a specific sequence of base pairs in DNA that determine the structure, function, and type of proteins that are made by the human body. Fusion genes play a critical role in many aspects of genomics and oncology. A fusion gene is the joining of two previously separate genes, ultimately forming a new combination of gene products. Learning mechanisms and techniques that can identify fusion genes accurately propose a way to use these fusion genes as a diagnostic and prognostic marker of cancer. Moreover, the fusion gene plays a critical role in cancer due to its tumor-specific expression. Fusion genes have been discovered in a wide diversity of cancers including sarcomas and carcinomas, and there have been over 2,000 genes discovered to be fusion genes. The first chromosomal rearrangement event that was found to be associated with cancer development was the discovery of the Philadelphia chromosome in 1960. The Philadelphia chromosome was an abnormally small chromosome that resulted from a translocation event between chromosomes 9 and 22 and was found to be present in approximately 95% of all cases of chronic myelogenous leukemia. In particular, this translocation event resulted in the fusion of the breakpoint cluster region (BCL) gene with the second exon of the Abelson murine leukemia viral oncogene homolog 1 (ABL1) gene.

5.1.2 MECHANISMS THAT GENERATE NEW FUSION GENES

The three major molecular events that can give rise to the generation of a new fusion gene include chromosome translocations, interstitial deletions, and chromosomal inversions (Figure 5.1). Chromosomal translocations are an abnormal rearrangement event that occurs when a fragment of one chromosome is transferred to another non-homologous chromosome and may be either classified as

Role of Transcriptomics in Identifying Fusion Genes and Chimeric RNAs

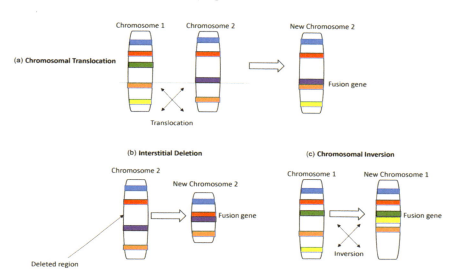

FIGURE 5.1 Overview of the three predominant events that can yield new fusion genes including translocations (a), interstitial deletions (b), and chromosomal inversions (c).

a reciprocal translocation or Robertsonian translocation. In reciprocal translocation, a portion of two non-homologous chromosomes is exchanged between chromosomes. These fragments may be of equal length or may be of varying sizes. In contrast, Robertsonian translocation occurs when the long arms of two non-homologous chromosomes attach to and stick to one another and occurs among the five acrocentric chromosomes including chromosomes 13, 14, 15, 22, and 23. Acrocentric chromosomes refer to chromosomes that have relatively low arm length ratios with somewhat small short arms and large long arms. Translocations may also be classified as either balanced, which occurs when an equivalent exchange of genetic material between two non-homologous chromosomes takes place, or unbalanced, which occurs when an unequal portion of genetic information is exchanged between two chromosomes and results in both missing and extra gene copies.

Interstitial deletions can create fusion genes when the intronic region between two genes is deleted, thus bringing the two formerly distant genes into close proximity. Finally, chromosomal inversions occur when a single chromosome undergoes some form of breakage and undergoes rearrangement and replication with itself, thus positioning two formerly distant genes close to each other.

5.1.3 Fusion RNA Transcripts

Chromosomal rearrangements can ultimately position distant genes into extremely close proximity to each other. When the transcriptional machinery is assembled at a given promoter region, this promoter may now become a shared promoter that allows for a long read-through transcript to be generated. This read-through transcript can then be translated into a functional fusion protein product. Polymerase read-through creates a chimeric transcript between two adjoining genes. Polymerase read-throughs transpire in conjunction with exons of adjoined genes, most prominently between the next-to-last exon of the upstream gene and the second exon of the downstream gene; consequently, new proteins are produced with domains from the two parent genes. These events ultimately ensure the heterogeneity of the proteome.

However, not all fusion transcripts and their resulting fusion proteins are generated as a result of chromosomal rearrangements. Instead, fusion RNA transcripts may arise due to one of two aberrant splicing mechanisms that may occur even in the absence of rearrangement events (Figure 5.2). First, trans-splicing occurs when exons from two or more different RNA transcripts are spliced together, resulting in fusion RNAs with repetitive exons, shuffled exon ordering, or exons that are transcribed from different strands. The second aberrant form of splicing that may produce oncogenic fusion transcripts that promote cancer progression is termed cis-splicing, which occurs when a single precursor mRNA is transcribed from two neighboring genes on the same chromosome and subsequent splicing mixes exons from both genes in order to produce novel fusion transcripts. In some instances, chromosomal rearrangements and aberrant splicing mechanisms can give rise to the same fusion transcript in cancer genotypes. For example, both trans-splicing and rearrangement events have been shown to generate a JAZF1–JJAZ1 fusion transcript that is composed of exons 1–3 derived from JAZF1 and the last 15 exons of JJAZ1 in endometrial stromal tumors.

Oncogenic chimeric transcripts usually fall into one of the following three groups based on the constituent genes comprising the chimeric RNAs:

1. Tyrosine kinase fusions that are constitutively active.
2. Serine-threonine kinase fusions that are constitutively active.
3. Chimeric transcripts encoding transcription factors that can lead to dysregulated expression (overexpression or decreased expression) of target genes.

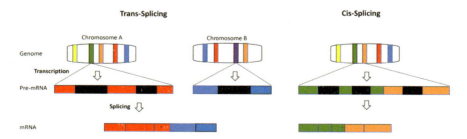

FIGURE 5.2 Overview of the trans-splicing and cis-splicing mechanisms that may generate oncogenic fusion transcripts.

Role of Transcriptomics in Identifying Fusion Genes and Chimeric RNAs   **133**

In many instances, fusion transcripts involving kinases will often gain the means to be constitutively activated even in the absence of ligand binding. This state of ligand-independent constitutive activation can result in the upregulation of downstream signaling cascades that promote the development and progression of tumors. Tyrosine kinase chimeric transcripts will frequently contain sequences corresponding to such genes as *ALK*, *NTRK1*, and *RET*, which have been found in many different cancer phenotypes including carcinomas of the breast, neck, prostate, and lungs. Serine-threonine kinase fusion transcripts containing such kinases as *AKT1* and various cyclin-dependent kinases have also been discovered in multiple tumor cells. Typically, kinase chimeras of this class upregulate downstream signaling pathways that increase cell growth, promote cell survival, and provide a proliferative advantage by suppressing the activation of apoptotic pathways.

Finally, chimeric proteins composed of transcription factors may result in the synthesis of proteins that either constitutively upregulate target genes or interact with new target genes that differ from proteins that the non-chimeric counterparts would normally associate with. For example, the MPRSS2–ERG fusion protein, which is the most commonly found fusion protein in prostate-derived tumors, promotes the overexpression of many transcription factors belonging to the E26 transformation-specific (EST) family, resulting in increased cell proliferation and angiogenesis.

5.1.4 The Connection between Fusion Genes and Non-Coding RNAs

The ncRNA expression profile of a given cancer has shown promising results with regard to prognostic and diagnostic information. As previously stated, non-coding RNA clusters frequently exist in fragile areas of the genome where replication-induced breakage and translocations are most likely to occur. Multiple cytogenetic aberrations and translocations have been shown to be associated with dysregulated ncRNA expression profiles. For example, the miRNA species *miR-196* was shown to be significantly downregulated in leukemia patients with 11q23 translocations. Another instance of this close connection between the presence of chromosomal translocations and dysregulated miRNA expression is *miR-21*, which was more highly expressed in leukemia patients with a t(6;11) than leukemia patients expressing an abnormal t(9;11).

Systematic expression analysis of lncRNAs has also demonstrated the possible usefulness of lncRNAs in the prognosis and diagnosis of various cancer subtypes. For example, a lncRNA produced from the *LOC100289656* pseudogene was highly expressed in leukemias with rearrangements involving the mixed-lineage leukemia (*MLL*) gene yet was silenced in leukemia cases lacking the MLL translocation. Thus, the expression of both miRNAs and lncRNA species is significantly linked to whether or not chromosomal translocations have occurred and may play a significant role in the severity and progression of cancer, particularly blood-borne cancers.

In previous sections, we discussed how most fusion genes are derived from constitutively activated kinases or transcription factors. In many subtypes of leukemia, the predominant type of fusion genes detected are ones produced from translocations involving genes encoding for transcription factors. Dysregulated ncRNA expression profiles may be induced directly by fusion proteins resulting from translocations through one of four possible mechanisms:

1. Interaction between fusion proteins and regulatory sequences for ncRNAs.
2. Downstream effectors of a fusion protein may mediate ncRNA dysregulation through transcriptional regulation.
3. Downstream effectors of a fusion protein may mediate ncRNA dysregulation through kinase activity.
4. Fusion proteins may disrupt RNA processing and formation.

First, chimeric proteins may directly bind to regulatory sequences that either upregulate the expression of ncRNAs or completely silence their expression. For instance, in leukemias with a t(11;16) event, the *MLL* gene fuses in the same reading frame as approximately 60 different genes. In addition to retaining its chromosome-binding domain in its N-terminal, common fusion protein partners include the transcriptional regulators AF4, AF9, ENL, and ELL. As an example, MLL-AF9 fusion proteins can bind to the promoter region for *miR-9* and lead to its overexpression by recruiting the histone H3 lysine 79 (H3K79) methylase DOT1L. To further corroborate the idea of direct regulation of ncRNA expression by fusion proteins, studies have shown enriched H3K79 methylation on the promoter regions of multiple lncRNAs that were upregulated in MLL-fusion leukemias.

In some instances, it is not the direct interaction with fusion proteins that elicits dysregulated ncRNA expression profiles but rather, downstream effectors of these fusion proteins that dictate whether or not particular ncRNAs are expressed. For example, HOXA9, which is a notable downstream effector for MLL-fusion proteins, has an antagonistic relationship with GIF1 and both compete for the same binding sites that correspond to *miR-196b* and *miR-21*. The expression of these miRNAs is largely dependent on the binding of HOXA9 to their promoter region, which is suppressed upon GIF1 binding. Thus, MLL-fusion proteins promote the overexpression of the effector HOXA9 and subsequently, through indirect means, increase the expression of *miR-196b* and *miR-21*.

The third mechanism that fusion proteins can mediate ncRNA dysregulation through is by modulating the activity of downstream effector proteins with inherent kinase activity. The fusion protein BCR-ABL has been shown to promote the upregulation of *miR-17–92* and lncRNA H19 through the phosphorylation of the downstream target c-Myc in chronic myelogenous leukemia cells. In addition, ablation of BCR-ABL activity by the BCR-ABL inhibitor imatinib reversed this aberrant ncRNA expression profile and restored the expression of *miR-17-92* and lncRNA H19 to physiologically normal levels.

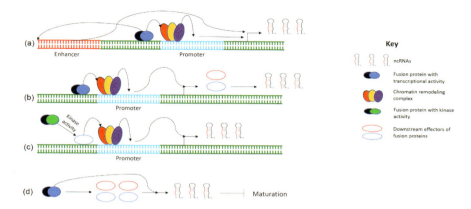

FIGURE 5.3 Overview of the four mechanisms in which chimera proteins may dysregulate the expression patterns of ncRNAs. (a) Fusion proteins with transcriptional activity may either interact with enhancer sequences or may recruit chromatin remodeling complexes in order to upregulate target ncRNAs. (b) Fusion proteins with transcriptional activity may recruit chromatin remodeling complexes to the promoter regions of downstream effectors that dysregulate ncRNA expression. (c) Fusion proteins with kinase activity directly activate downstream effectors that then recruit chromatin remodeling complexes to the promoters of target ncRNAs. (d) Fusion proteins with transcriptional activity inhibit the maturation of ncRNAs, either directly or indirectly through downstream effector proteins.

Finally, some studies have shown that chromosomal rearrangements, and the resulting functional fusion proteins, can promote dysregulated expression of ncRNAs by disrupting the RNA processing machinery. One study demonstrated that MLL-fusion proteins as well as activated c-Myc can bind to and activate the promoter region of *miR-150*, resulting in the synthesis of nascent pri-miRNA molecules. However, a secondary target that is upregulated by c-MYC is the lncRNA Lin28, which blocks the maturation of pri-miRNAs. Thus, there is an upregulation of the nonfunctional pri-miRNAs but a decrease in the expression of mature forms of *miR-150*, ultimately resulting in suppressed *miR-150* function. Figure 5.3 shows an overview of the four mechanisms in which chimera proteins may promote a dysregulated ncRNA environment.

5.2 DETECTION METHODS FOR IDENTIFICATION OF FUSION GENES AND CHIMERIC PROTEINS

5.2.1 Guided Detection Approaches

The first fusion gene, BCR-ABL, was discovered in blood-borne cancers using chromosome banding techniques, which encompass staining methodologies that can be utilized to differentiate between normal and abnormal chromosomes with translocation. These techniques allow for the identification of chromosomal regions based on unique banding patterns. The development of more sophisticated methods such as fluorescence in situ hybridization, polymerase chain reaction, and comparative genomic hybridization enabled the identification of fusion genes now in solid tumors as well as hematologic cancers.

Fluorescence in situ hybridization uses fluorescent probes to aid in the visualization of chromosome structures and breakpoints in nondividing cells as well as chromosomes with rearrangements during metaphase. Comparative genomic hybridization assays use fluorescent probes to compare the copy number variations of chromosomes between a test sample and a reference sample in scenarios where unbalanced chromosomal abnormalities have occurred. However, these guided approaches cannot be used to identify fusion transcripts generated by trans-splicing and cis-splicing events because no alterations in chromosomal structure have occurred. Instead, we can use techniques such as reverse transcription polymerase chain reaction and RNase protection assays, the latter of which uses fluorescently labeled probes that are complementary to an mRNA sequence of interest and can be used to identify a particular RNA from a heterogeneous mixture.

The advent of high-throughput sequencing technologies allows us to now identify fusions at both a DNA and an RNA level without prior knowledge as to the cytogenetic characteristics of the tumor in question. Subsequent sections will discuss methods and tools that use high-throughput sequencing data to identify fusion genes and chimeric transcripts.

5.2.2 High-Throughput Sequencing-Based Detection Methods

There are multiple algorithms that may be used for the detection of fusion genes and the detection of transcription-induced chimeras using RNA-sequencing methods. In general, methods based on RNA-sequencing data include the following three essential steps:

1. Filtering of chimeric reads.
2. Junction detection.
3. Chimeric transcript assembly.

RNA-seq uses a mapper for chimeric transcript detection, which leads to two different reads: (1) discordant paired-end reads such as those on different chromosomes and (2) "split" reads. These lead to multiple different approaches including

1. paired-end+fragmentation approaches such as ChimPipe,
2. direct fragmentation approaches such as GFusion,
3. whole paired-end approaches such as InFusion and STAR-Fusion.

We will first discuss the paired-end+fragmentation methodology and ChimPipe tool in the next section.

Role of Transcriptomics in Identifying Fusion Genes and Chimeric RNAs

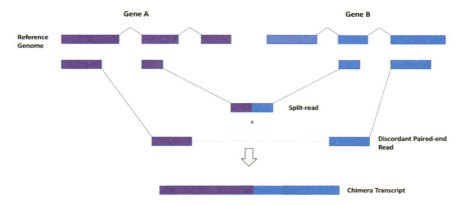

FIGURE 5.4 The distinction between discordant paired-end reads and split reads as well as how they are combined to detect chimeric transcripts is presented.

5.2.3 CHIMPIPE

ChimPipe is a method used for the primary identification of fusion genes and chimeras from paired-end Illumina RNA-seq data using the paired-end+fragmentation method. Many programs that employ the paired-end+fragmentation method first find discordant paired-end reads, construct an exon–exon junction database, and then map the currently unmapped reads using the exon–exon junction database. Thus, these platforms cannot identify split reads that do not have a corresponding discordant paired-end read. Split reads occur when a particular sequence of a read maps to one area of the reference genome and the other portion of the read maps to a different area of the reference genome. In contrast to this stepwise process, ChimPipe allows for the identification of these reads independently of one another and can identify chimeric junctions while simultaneously controlling the false positive rate. Illustrations of a discordant paired-end read and a split read are presented in Figure 5.4. The ChimPipe tool operates using the following four-step process.

5.2.3.1 Exhaustive Paired-End and SPLIT Read Mapping with Genome Multitool (GEM)

The paired-end reads are first mapped in three different ways using the GEM RNA-seq pipeline. The reads are mapped to the genome, then to the transcriptome (consisting of combinations of exons in each gene), and lastly de novo splice junctions. In regard to de novo splice junctions, reads are split into two different junctions of a minimum of 15 base pair length and mapped to the genome. These split-mappings must be less than 4% of all mismatches to ensure a low false positive rate. In order to emphasize a higher mapping sensitivity, a subsequent attempt is utilized by deteriorating only two base pairs at the terminal of each segment. At this point, segments can only map to different positions on the same chromosomes. Any pairs that map to at least ten different positions are denoted as being unmapped. Lastly, GEM-based methods are used to map unmapped reads

138   RNA-seq in Drug Discovery and Development

in a second de novo mapping using a split-mapped method to find splice junctions that connect loci located on different chromosomes.

5.2.3.2 ChimSplice

Split-mapped reads are subsequently made into an agglomeration of reads extending the same splice junction and ensuring a 5'–3' orientation, which is essential in the determination of upstream and downstream parent genes. *ChimSplice* makes a consensus splice junction due to the utilization of exact junction coordinates, upstream coordinates, and downstream coordinates in their respective groups. Staggered split reads, which refer to reads that span across the same splice junctions yet map to unique starting points on both sides of the junction, are also present which help to lower the presence of false positives due to their specificity to authentic chimeric transcripts. Finally, splice junctions are then selected for analysis.

5.2.3.3 ChimPE

In the previous step, *ChimSplice* utilized the split reads to map to the genome and identify splice possible junctions. In this step, paired-end reads generated from all three alignment steps including genome, transcriptome, and de novo alignments are then utilized to corroborate the possible splice junctions predicted by the split reads. Only paired-end reads that have both mates mapping to exons from at least two different genes are considered and any other paired-end reads are filtered out during this step. For each chimeric junction, discordant paired-end reads connecting their parent genes are selected and their mapping position is subsequently evaluated. Discordant paired-end reads that further support the existence of a splice junction are termed consistent paired-end reads while reads that contradict the existence of a possible splice junction are termed inconsistent paired-end reads. Inconsistent paired-end reads may arise in multiple instances due to alternative chimeric RNA isoforms, misalignment of the paired-end reads, or are a false positive identified by the ChimPipe algorithm.

5.2.3.4 ChimFilter

Chimeric junction candidates are lastly filtered in order to ensure reliable and accurate chimeras. The underlying assumption that governs this step of the ChimPipe algorithm is that false positive splice junctions that arise due to misalignments will not be corroborated by both split reads and the discordant paired-end reads. Chimeric junctions with annotated splice sites must be supported by at least one consistent paired-end read, one split read, and three total reads; novel splice sites need the support for a minimum of three paired-end reads, three split reads, and six total reads. Moreover, chimeric junctions involving genes on either the mitochondrial chromosome or are pseudogenic are filtered out. In addition, other junctions filtered out include chimeras from genes that have a high sequence identity (at least 90%). These filtrations account for potential misalignments or mapping errors. ChimPipe simply requires a genome and gene annotation as well as a pair of RNA-seq fastq files.

5.2.4 GFUSION

Another effective algorithm using data files derived from RNA-sequencing in order to accurately detect fusion genes is a method called GFusion. GFusion is a method in which RNA-seq data is used to identify fusion genes that may play a critical role in cancer initiation. GFusion is a software made to identify fusion genes from paired-end RNA-seq read data and from single-end RNA-seq data. These are first aligned against a reference genome using a splice-aware alignment tool such as TOPHAT or HISAT2 to map our raw reads. This is followed by a series of filtering steps in order to identify the strongest candidates for non-artifactual chimeric transcripts. In this pipeline, a fusion boundary is defined as the exact genomic position in which the breakpoint between two genes is located. A split read is defined as a sequence read directly containing a fusion boundary while a split fragment refers to a paired-end read where one of the reads is a split read and the other mate read has a unique alignment that is dependent on the split read. Finally, a spanning fragment is defined as a paired-end read that contains a fusion boundary with two reads aligning to two different genes. A graphical representation of these terms is presented in Figure 5.5 while an overview of the GFusion algorithm is presented in Figure 5.6.

5.2.4.1 The Beginning Alignment

GFusion finds fusion genes due to their alignment locations of reads. In order to best identify this, GFusion uses basic splice-aware alignment tools. To begin the GFusion process, all reads must be aligned to the reference genome. TopHat is used as the initial alignment tool in the reference manuscript although any splice-aware alignment tool could be used. TopHat ultimately produces sequence alignment map (SAM) files by utilizing Bowtie to differentiate between mappable and

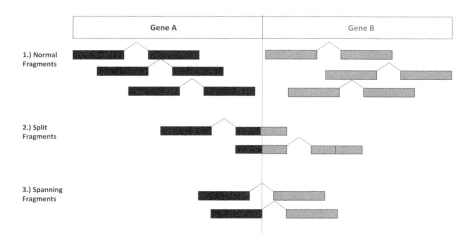

FIGURE 5.5 Visual representation of the key definitions employed by the GFusion pipeline.

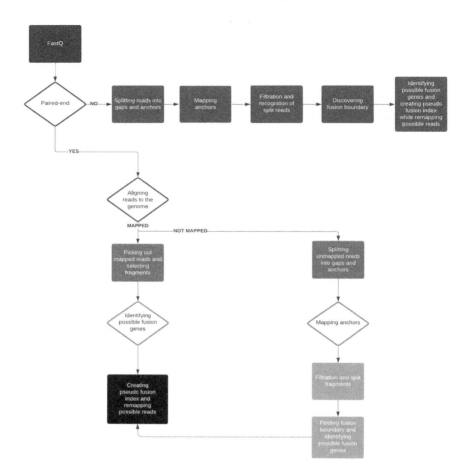

FIGURE 5.6 Overview of the steps utilized by the GFusion pipeline to identify fusion genes from either single-end reads or paired-end reads.

unmappable reads. By default, the GFusion pipeline filters out all nonfunctional unmapped reads with a mapping quality score of ≤ 30.

5.2.4.2 Anchors

GFusion accurately splits unmapped reads into three shorter segments. Its mate reads are aligned to the genome with high mapping quality. The middle segment is a gap whose gap length is equivalent to the read length minus the anchor length. The initial and terminal segments are defined as "anchor_1 and anchor_2" with 20 bp or greater length, which must not exceed 50% of the length of a given read (default = 40%). In addition, the query name and the quality score for each anchor are carried over from the original parental reads. With these conditions, the two

Role of Transcriptomics in Identifying Fusion Genes and Chimeric RNAs **141**

anchors as well as the gap make special paired-read reads where the anchors are the paired-end reads and the gap is the insert segment.

5.2.4.3 Alignment and Localization

Anchors are next aligned to the reference genome using Bowtie. GFusion next identifies potential split reads with two anchors aligning to different genes while simultaneously discarding reads in which both anchors map to the same gene. Paired-end reads in which anchor_2 of a split read as well as the mate read map to the same gene will ultimately be considered potential split fragments. GFusion is unique in the fact that it considers the mate reads and split reads alignment results. This is evident in the observed lower rate of false positives.

5.2.4.4 Fusion Boundaries

Fusion boundaries are located by satisfying specific criteria using the human gene annotation as well as potential split reads that harbor the fusion boundaries. Potential split reads with alignments where two anchors located near the exon boundaries are chosen, and exon junction boundaries are used as an approximation of the fusion boundaries. The below condition is what the most probable range of anchor location is

$$p < Cor - G_1 < gl + p, \quad \text{if anchor}_1 \text{ strand} - \text{or anchor}_2 \text{ strand} +$$

$$p < G_2 - Cor < al + gl + p, \quad \text{if anchor}_1 \text{ strand} + \text{or anchor}_2 \text{ strand}$$

where Cor denotes the coordinate of the first base in the alignment, strand+ refers to anchors that align to the forward reference sequence, strand− refers to anchors that align to the reverse-complement reference sequence, $G1$ denotes the starting position for a given exon, $G2$ denotes the ending coordinate of a given exon, al denotes the anchor length, gl refers to the length of the gap, and p indicates the maximum allowed distance (the difference) between the boundary site of one anchor and the mapped exon boundary site.

5.2.4.5 Confirming Fusion Models

There are four possible fusion models for fusion gene strand direction. *ff* is one fusion model in which downstream and upstream genes are forward transcribed and spliced in the same orientation. The *rr* model contains fusion genes that both have reversed strand orientations while *fr* is another fusion model which is composed of a downstream gene with reverse orientation and an upstream gene with forward orientation. Lastly, the *rf* model links the forward strand of the downstream gene to the reversed strand of the upstream gene.

5.2.4.6 Grouping Candidate Fusions

GFusion identifies candidate fusions specifically on the number of split and spanning fragments, the latter of which spanning fragments harbor fusion boundaries in the insert segments. Each read aligns to different genes. Initially, aligned reads

142 RNA-seq in Drug Discovery and Development

with mapping scores over 30 are selected. GFusion removes false positive fusions by only allowing fusion candidates supported by one spanning fragment and one split fragment, at least.

5.2.4.7 Fusion Index and Realignment

This step concludes the filtration process to eliminate false positives. The Bowtie reference index is used with the potential fusion sequences. Upstream and downstream sequences are joined to make potential fusion reference sequences at boundaries. For instance, if the applicable fusion model is *fr*, the GFusion pipeline will define the upstream sequence ranging from the first base of the read to the base of the fusion boundary as well as define the downstream sequence starting with the base of the fusion boundary to the final base of the downstream gene. Each split fragment and spanning fragments are realigned against the index to recount the number of supporting fragments.

5.2.5 INFUSION

InFusion is an additional computational method that can be utilized to identify and discover chimeric transcripts. In particular, the InFusion pipeline is unique in that it can identify alternatively spliced chimeric transcripts that may be contained within intergenic regions of the genome. The algorithm for InFusion reconstructs likely fusion transcripts by detecting both SPLIT reads, which refer to sequence reads that span fusion junctions, and BRIDGE reads, which refer to paired-end reads in which a pair of reads derived from the same fragment cross the fusion boundary located within the insert.

Assuming paired-end reads are used as input into the pipeline, InFusion detects chimeric transcripts in a seven-step procedure and is summarized in Figure 5.7. Briefly, InFusion begins with the alignment of reads to both a reference genome and a transcriptome. During this step, any unmapped reads are pooled together and aligned locally to the genome. Next, any locally aligned mapped reads are used to identify possible SPLIT reads that span fusion junctions. Simultaneously, insert size statistics are collected and BRIDGE read pairs are derived from discordantly mapped read pairs. Any SPLIT and BRIDGE reads that are designated as possibly belonging to the same fusion gene are clustered together. Finally, the library of possible fusion genes is filtered in order to remove possible false positive results and the resulting true positive chimeric transcripts are reported. Each step will next be discussed in further detail.

5.2.5.1 Short Reads Alignment

The first step of the InFusion pipeline is to align our raw reads to a reference genome and transcriptome in order to identify possible fusion genes. Ultimately, this step is concerned with differentiating between short reads that map to putative fusion genes and short reads that map to non-fusion or genetically normal genes. In addition, read alignments collect statistical information concerning the sample distribution of insert sizes as well as expression levels in order to aid in

Role of Transcriptomics in Identifying Fusion Genes and Chimeric RNAs   143

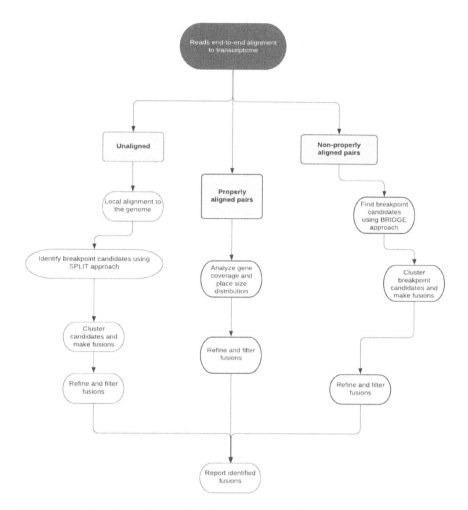

FIGURE 5.7 Outline depicting all of the required steps for the identification and reporting of fusion genes by the InFusion pipeline is presented. Note the distinct steps initiated with unaligned, properly aligned pairs, and non-properly aligned pairs.

the filtering of false positives downstream in the pipeline. The default short read alignment tool utilized in the InFusion pipeline is Bowtie2; however, other short read aligners such as Subread, GEM, and BWA may be used as well.

5.2.5.2   Local Short Reads Alignment

In order to find probable SPLIT read candidates, unaligned reads are going to be locally aligned to the genome in this step using Bowtie2 in "local mode". InFusion uses special parameters to validify mapping in terms of local alignment. These special parameters can be further expressed by the equation (Eq. 5.1):

$$\frac{Score_{\max}}{3} \qquad (5.1)$$

Using an equation, InFusion can validify mapping with scores greater than the set equation. $Score_{\max}$ denotes the maximum alignment score for a specific scoring scheme. The default scoring scheme utilized by InFusion is as follows: -2 for mismatch, 2 for match, -6 for gap open, and -3 for gap extension.

5.2.5.3 Local Alignment Analysis

The following conditions are defined in order to analyze the local alignments and assess possible SPLIT read candidacy:

$$\{A_1, A_2, A_3 \ldots \ldots, A_n\}$$

where A_1 denotes a specific sequence (chromosome) c_i, p_l represents the starting coordinates of the reference sequence, q_l denotes the starting point in the read coordinates, and l_i denotes the length. This is shown by the function:

$$A_1\left[c_i, p_i, q_i, l_i\right]$$

Two local alignments will form a SPLIT read as long as two principles hold true:

1. Either c_i is NOT equal to c_j or $p_i + l_i - p_j > l_{\max}$
   This principle indicates that the alignments are essentially derived from different chromosomes or the distance between these alignments is greater than the maximum intron size. By default, the maximum intron length, $l_{\max}$, is 20,000 bp for human genomes.
2. $q_i < q_j$ and $(1) \times T_{outer} < q_j - (q_j + I_i) < T_{inner}$
   This principle holds true when the alignments are concordant in the read coordinates based on maximum intersection size, T_{inner}, as well as the maximum distance between alignments T_{outer} in the coordinates by the read. The default settings for T_{inner} and T_{outer} are 10 and 2 bps, respectively.

5.2.5.4 Analysis of End-to-End Alignments

In this step, non-properly aligned pairs are analyzed, in which the transcriptomic alignments are converted to genomic coordinates. The alignments *M1* and *M2* form a properly aligned pair if they are aligned on the same chromosome within a distance of the defined maximum intron size $I_{\max}$. Otherwise, the alignments *M1* and *M2* form a possible BRIDGE read pair. All the identified BRIDGE pairs are then subsequently analyzed, and the distribution of insert size and expression levels is estimated for each gene.

5.2.5.5 Clustering and Establishing Putative Fusions

The clustering method used categorizes the alignments which make the breakpoint candidates into clusters due to their genomic coordinates. The alignments

Role of Transcriptomics in Identifying Fusion Genes and Chimeric RNAs 145

are analyzed on multiple factors. One such factor of analysis is that the alignment makes a new cluster only if there is no intersection with other clusters. If there is an intersection with another cluster, the alignment is added to this specific cluster. Lastly, if the alignment is associated to multiple clusters at the same time, the clusters are initially meshed into one single cluster, and then the alignment is added. Of note, SPLIT reads suggest an exact coordinate for the breakpoint while a BRIDGE read pair implies an approximate genomic coordinate for the breakpoint between two genes. These clusters can contain a variety of breakpoints due to this clustering procedure. The next step is to separate the clusters solely based on directionality of the cluster, which is then proceeded by an analysis of whether the coordinates of the breakpoint candidates in a cluster are compatible within a specific tolerance (default = 10 bp). If this does not occur, then the cluster is split into two new clusters. This continues until there is no difference in the coordinates.

5.2.5.6 Purifying and Filtering Fusions

When utilizing paired-end reads, the SPLIT reads and the BRIDGE reads will be in agreement as to the identity of a fusion gene. Thus, one filtering step employed during this step is to remove SPLIT reads that are not congruent with the BRIDGE reads for putative fusion genes. Reads that are mapped to multiple fusions are assigned to the putative fusion with the highest assigned score that is calculated using the alignment type, multi-mapping status, and whether or not a mate pair is present alongside SPLIT reads. Additional features associated with each putative fusion are calculated to further filter candidates including meta cluster homogeneity, read coverage, homology, insert size, strand specificity, and biological type.

5.2.5.7 Reporting Fusions

The output file from the InFusion pipeline contains information reports concerning the associated genomic regions, the breakpoint coordinates, the number of supporting SPLIT and BRIDGE reads for each fusion gene, and the features computed in the previous step.

5.2.6 STAR-FUSION

STAR-Fusion was developed as part of the Trinity Cancer Transcriptome Analysis Toolkit (CTAT) and was designed to provide bioinformatics methods to translate RNA-seq data into cancer-related insights such as cancer transcriptomes, identification of fusion genes, and the detection of driver mutations. Figure 5.8 provides an overview of the STAR-Fusion pipeline. The STAR-Fusion pipeline requires Illumina RNA-seq data as input and generates an output file containing all of the identified split reads and discordant reads that correspond to chimeric transcripts.

STAR-Fusion utilizes the previously discussed STAR alignment tool to map raw reads to exonic features contained in the precompiled metadata library provided by CTAT based on overlapping coordinates. The first step in the STAR-Fusion pipeline filters out read alignments between gene pairs that are localized to regions of comparable sequences. Candidate fusion genes are queried against

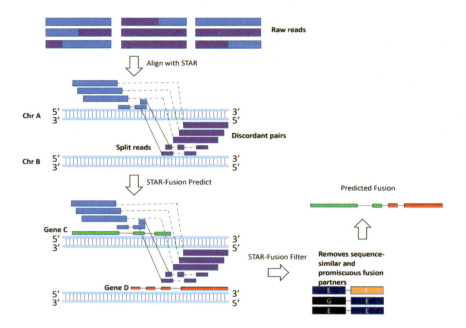

FIGURE 5.8 Overview of the STAR-Fusion pipeline.

an all-vs-all *blastn* database in order to identify regions with a high degree of sequence similarity. Read alignments for candidate chimeras that have significant overlap with the reference sequences are removed and duplicate reads are also filtered out of the pool of possible chimeras. The alignments that remain after these filtering steps are then assigned to preliminary fusion gene pair candidates, which are given a score corresponding to the number of split reads that support the fusion breakpoint as well as the number of paired-end fragments that span the predicted fusion junction.

Once the list of preliminary fusion gene candidates has been compiled, this list undergoes two additional filtering stages. The first step is the basic filtering process in which the strength of the evidence in support of the fusion gene candidate being a true fusion gene, and thus not a false positive, is assessed. Each fusion gene must have at least two RNA-seq fragments that support the presence of a true fusion gene, one of which must be a split read that delineates the location of the fusion breakpoint. In cases where the fusion breakpoint is novel and does not correspond to known exon splice sites denoted in the annotated reference, then at least three split reads in support of the breakpoint's location are required. If the candidate fusion gene is only supported by split reads and lacks fragments that span the fusion junction, at least 25 bases both upstream and downstream of the splice junction must have reads that align to them. During the advanced fusion filtering process, the following operations are performed that assess fusion characteristics from the perspective of both the individual fusion pair and through its comparison with other called fusion gene predictions in our sample:

Role of Transcriptomics in Identifying Fusion Genes and Chimeric RNAs 147

1. **Fusion paralog filter**: This step excludes any fusion candidate, denoted GeneA-GeneB, in which both genes are paralogs of one another. In addition, if a candidate fusion gene, denoted GeneA-GeneC, exists in which the probability of GeneC being a paralog of GeneB is high, and if more fusion evidence supports the existence of the GeneA-GeneB than the existence of GeneA-GeneC, the predicted GeneA-GeneC fusion will be removed.

2. **Promiscuous fusion filter**: If multiple alternative fusion candidates are detected on the list including GeneA-GeneB, GeneA-GeneC, and GeneB-GeneC, and if the evidence in favor of being a true fusion gene is substantially higher for the candidate GeneA-GeneB, all other alternative candidates will be discarded. In instances where GeneA is identified as comprising possible fusion candidates with ≥ 10 unique partners, all fusion pairs containing GeneA will subsequently be filtered out.

3. **Red herring filter**: Fusion pairs are next annotated using the FusionAnnotator module and the CTAT Human Fusion library database. Annotated fusion pairs that have been found in normal RNA-seq datasets lacking fusion proteins, including having a mitochondrial or human leukocyte antigen partner, are removed. In addition, any gene pairs composed of two immunoglobulin gene segments are also discarded.

4. **Fusion expression filter**: The abundance of RNA-seq fragments that support the existence of a given fusion gene is normalized by sequencing depth and is represented by fusion fragments per million total RNA-seq fragments (FFPM). Any remaining fusion candidates with an FFPM of ≤ 0.1 are discarded.

In order to generate a sample data output using the STAR-Fusion pipeline, we will adapt a github protocol using its sample data inputs. The source folder containing our required inputs can be downloaded directly from the following link: https://github.com/STAR-Fusion/STAR-Fusion-Tutorial/releases/tag/v0.0.1. In particular, we will need to upload the following files into our Galaxy history:

1. **Minigenome.fa**: A small genome sequence comprising 750 genes.
2. **Minigenome.gtf**: Provides the structural transcript annotations for the reference genes.
3. **Rnaseq_1.fastq.gz**: Corresponds to the "left reads" of a paired-end dataset.
4. **Rnaseq_2.fastq.gz**: Corresponds to the "right reads" of a paired-end dataset.

Once we have successfully downloaded the files to our local computer and subsequently uploaded them to our Galaxy history, we now have all of the requirements needed to run the STAR-Fusion tool. The following parameters (Figure 5.9) were utilized to generate the data output:

Set *use output from earlier STAR run or let STAR-Fusion control running STAR* to let STAR-Fusion control running STAR. This will tell the STAR-Fusion tool to first use STAR to align our paired-end reads to the reference genome. An additional option is to run STAR on our reads separately and use the output from that run as input for this option in STAR-Fusion.

1. Set *left.fq file* to our rnaseq_1.fastq.gz file.
2. Set *right.fq file* to our rnaseq_2.fastq.gz file.
3. Set *source for sequence to search* to sequences from your history.
4. Set *select the reference genome (FASTA file)* to our minigenome.fa file.
5. Set *gene model (gff3, gtf) file for splice junctions and fusion gene detection* to our minigenome.gtf file.
6. Set *result of BLAST+-blastn of the reference fasta sequence with itself* to minigenome.fa (as tabular). This file should be automatically selected as input for this option and will not be directly seen in the Galaxy history.
7. Set *settings to use* to use defaults.

Completion of the STAR-Fusion job will generate a tabular output (Figure 5.10) with the list of all fusion genes detected by the tool and are ordered from the strongest support to the weakest support (determined by the number of junction reads and spanning fragments that support the existence of the fusion gene). The output file also lists the left and right breakpoints of the two genes comprising the fusion product as well as the name of each supporting junction read and spanning fragment. However, due to the size of the resulting file, Figure 5.10 will not include the names of supporting spanning fragments.

FIGURE 5.9 Parameters used to generate the STAR-Fusion output.

#fusion_name	JunctionReads	SpanningFrags	Splice_type	LeftGene	LeftBreakpoint	RightGene	RightBreakpoint	JunctionReads
TATDN1--GSDMB	31	126	ONLY_REF_SPLICE	TATDN1^TATDN1	minigenome:6584842:-	GSDMB^GSDMB	minigenome:2142920:-	fustut00000342d7,fustut00001ab157,fustut00001b722d,fustut00001
TATDN1--GSDMB	31	126	ONLY_REF_SPLICE	TATDN1^TATDN1	minigenome:6584842:-	GSDMB^GSDMB	minigenome:2138981:-	fustut00001464fd,fustut0000028b25,fustut00001464fc,fustut00000b
ACACA--STAC2	32	41	ONLY_REF_SPLICE	ACACA^ACACA	minigenome:1869270:-	STAC2^STAC2	minigenome:2067875:-	fustut000011f133,fustut00001a4353,fustut0000170489,fustut000013
CCDC6--RET	27	14	ONLY_REF_SPLICE	CCDC6^CCDC6	minigenome:609384:-	RET^RET	minigenome:490475:+	fustut000010712e,fustut0000072739,fustut00000b2fe0,fustut00000
BCAS4--BCAS3	8	81	ONLY_REF_SPLICE	BCAS4^BCAS4	minigenome:4187310:+	BCAS3^BCAS3	minigenome:2490128:+	fustut00000c3636,fustut00000bd70c,fustut0000157aa9,fustut00001
BCAS4--BCAS3	4	81	ONLY_REF_SPLICE	BCAS4^BCAS4	minigenome:4187310:+	BCAS3^BCAS3	minigenome:2485565:+	fustut0000121d67,fustut00000ada6b,fustut00000bbf73,fustut00001
RPS6KB1--SNF8	18	20	ONLY_REF_SPLICE	RPS6KB1^RPS6KB1	minigenome:2354713:+	SNF8^SNF8	minigenome:2265737:-	fustut00000d1ae8,fustut00000e17bc,fustut00000c1949,fustut00000
TOB1--SYNRG	17	23	ONLY_REF_SPLICE	TOB1^TOB1	minigenome:2296883:-	SYNRG^SYNRG	minigenome:1999668:-	fustut000000ed69,fustut000011098a,fustut0000026ea0,fustut00000
VAPB--IKZF3	10	34	ONLY_REF_SPLICE	VAPB^VAPB	minigenome:4256867:+	IKZF3^IKZF3	minigenome:2124385:-	fustut000017fb56,fustut00000927de,fustut00000b87ce,fustut00001
BAG6--SLC44A4	17	5	ONLY_REF_SPLICE	BAG6^BAG6	minigenome:5774709:-	SLC44A4^SLC44A4	minigenome:5786784:-	fustut00000457e2,fustut0000187f21,fustut00001475c8,fustut00001
BCR--ABL1	13	13	ONLY_REF_SPLICE	BCR^BCR	minigenome:4683927:+	ABL1^ABL1	minigenome:6820600:+	fustut00000f062f,fustut0000173c3a,fustut0000161612,fustut00000a
AHCTF1--NAAA	4	40	ONLY_REF_SPLICE	AHCTF1^AHCTF1	minigenome:452828:-	NAAA^NAAA	minigenome:5230273:-	fustut000009d2fe,fustut00000a71dd,fustut000019ed59,fustut00001
CYTH1--EIF3H	7	20	ONLY_REF_SPLICE	CYTH1^CYTH1	minigenome:2680251:-	EIF3H^EIF3H	minigenome:6546559:-	fustut00000b29ec,fustut00000bd1ee,fustut00000bc477,fustut00001
VAPB--IKZF3	2	34	ONLY_REF_SPLICE	VAPB^VAPB	minigenome:4256867:+	IKZF3^IKZF3	minigenome:2123804:-	fustut000017f7bb,fustut000011601c
RARA--PKIA	7	14	ONLY_REF_SPLICE	RARA^RARA	minigenome:2192052:+	PKIA^PKIA	minigenome:6502055:+	fustut000008cdda,fustut000008bae7,fustut000008bae8,fustut00000
RAB22A--MYO9B	7	13	ONLY_REF_SPLICE	RAB22A^RAB22A	minigenome:4235031:+	MYO9B^MYO9B	minigenome:3118079:+	fustut000004dac6,fustut000003e3ec,fustut000001eaa2,fustut00001
ZMYND8--CEP250	2	32	ONLY_REF_SPLICE	ZMYND8^ZMYND8	minigenome:3950250:-	CEP250^CEP250	minigenome:3836708:+	fustut000000f10c,fustut000013ba40
VAPB--IKZF3	1	34	ONLY_REF_SPLICE	VAPB^VAPB	minigenome:4256867:+	IKZF3^IKZF3	minigenome:2125233:-	fustut000011601d
SUMF1--LRRFIP2	4	14	ONLY_REF_SPLICE	SUMF1^SUMF1	minigenome:4767693:-	LRRFIP2^LRRFIP2	minigenome:4900118:-	fustut00000f21c0,fustut000016f1b0,fustut00001678b7,fustut000015
MED1--STXBP4	3	15	ONLY_REF_SPLICE	MED1^MED1	minigenome:2119007:-	STXBP4^STXBP4	minigenome:2345982:+	fustut00001409b4,fustut000019c6a3,fustut0000163639
MED1--ACSF2	5	7	ONLY_REF_SPLICE	MED1^MED1	minigenome:2107134:-	ACSF2^ACSF2	minigenome:2266230:+	fustut0000121baa,fustut0000197b0e,fustut0000068cc0,fustut00001
CYTH1--EIF3H	1	20	ONLY_REF_SPLICE	CYTH1^CYTH1	minigenome:2680251:-	EIF3H^EIF3H	minigenome:6546337:-	fustut00001a4dbf
ANKHD1--PCDH1	4	7	ONLY_REF_SPLICE	ANKHD1^ANKHD1	minigenome:5561254:+	PCDH1^PCDH1	minigenome:5680729:-	fustut0000186a7a,fustut0000122eb9,fustut000002db67,fustut00001
SUMF1--LRRFIP2	2	14	ONLY_REF_SPLICE	SUMF1^SUMF1	minigenome:4769089:-	LRRFIP2^LRRFIP2	minigenome:4900118:-	fustut000006f18a,fustut000007a95d
ARFGEF2--SULF2	2	14	ONLY_REF_SPLICE	ARFGEF2^ARFGEF2	minigenome:4071002:+	SULF2^SULF2	minigenome:4059796:-	fustut00001634fe,fustut000006ccdc
LL22NC03-86G7.1--PABPC1P5	5	0	INCL_NON_REF_SPLICE	LL22NC03-86G7.1^LL22NC03-86G7.1	minigenome:4639326:+	PABPC1P5^PABPC1P5	minigenome:8036146:+	fustut0000128980,fustut000012897f,fustut000008aae0,fustut00000
SUMF1--LRRFIP2	1	14	ONLY_REF_SPLICE	SUMF1^SUMF1	minigenome:4778610:-	LRRFIP2^LRRFIP2	minigenome:4900118:-	fustut0000134787
RARA--PKIA	1	14	ONLY_REF_SPLICE	RARA^RARA	minigenome:2192052:+	PKIA^PKIA	minigenome:6505927:+	fustut00001139cd

FIGURE 5.10 Sample output generated from the STAR-Fusion tool.

5.3 SUMMARY

In this chapter, we introduced fusion genes and the processes that may occur to generate novel fusion products. We introduced several tools to identify and discover fusion genes in RNA-seq samples including ChimPipe, InFusion, GFusion, and STAR-Fusion, the latter of which was used in the Galaxy interface to generate a sample output file.

KEYWORDS AND PHRASES

After reading this chapter, you should be able to demonstrate familiarity with the following words and phrases:

- Understand what a fusion gene is and the molecular mechanisms that can produce them.
- Understand what fusion transcripts are and the molecular mechanisms that can generate them.
- Understand how fusion genes and non-coding RNAs are related including the four primary mechanisms in which fusion genes can dysregulated non-coding RNA expression profiles.
- Understand the principles underlying ChimPipe, InFusion, and GFusion.
- Understand the principles of STAR-Fusion and how to generate output files successfully in Galaxy.

BIBLIOGRAPHY

Banavali, N. K. (2013, Jun 5). Partial base flipping is sufficient for strand slippage near DNA duplex termini. *J Am Chem Soc, 135*(22), 8274–8282. https://doi.org/10.1021/ja401573j

Bartonicek, N., Clark, M. B., Quek, X. C., Torpy, J. R., Pritchard, A. L., Maag, J. L. V., Gloss, B. S., Crawford, J., Taft, R. J., Hayward, N. K., Montgomery, G. W., Mattick, J. S., Mercer, T. R., & Dinger, M. E. (2017, Dec 28). Intergenic disease-associated regions are abundant in novel transcripts. *Genome Biol, 18*(1), 241. https://doi.org/10.1186/s13059-017-1363-3

Bartonicek, N., Maag, J. L., & Dinger, M. E. (2016, May 27). Long noncoding RNAs in cancer: mechanisms of action and technological advancements. *Mol Cancer, 15*(1), 43. https://doi.org/10.1186/s12943-016-0530-6

Czech, B., & Hannon, G. J. (2011, Jan). Small RNA sorting: Matchmaking for argonautes. *Nat Rev Genet, 12*(1), 19–31. https://doi.org/10.1038/nrg2916

Dhanoa, J. K., Sethi, R. S., Verma, R., Arora, J. S., & Mukhopadhyay, C. S. (2018). Long non-coding RNA: Its evolutionary relics and biological implications in mammals. *J Anim Sci Technol, 60*, 25. https://doi.org/10.1186/s40781-018-0183-7

Ding, X., Zhu, L., Ji, T., Zhang, X., Wang, F., Gan, S., Zhao, M., & Yang, H. (2014). Long intergenic non-coding RNAs (LincRNAs) identified by RNA-seq in breast cancer. *PLoS One, 9*(8), e103270. https://doi.org/10.1371/journal.pone.0103270

Edwards, P. A. (2010, Jan). Fusion genes and chromosome translocations in the common epithelial cancers. *J Pathol, 220*(2), 244–254. https://doi.org/10.1002/path.2632

Friedlander, M. R., Mackowiak, S. D., Li, N., Chen, W., & Rajewsky, N. (2012, Jan). miRDeep2 accurately identifies known and hundreds of novel microRNA genes in seven animal clades. *Nucleic Acids Res, 40*(1), 37–52. https://doi.org/10.1093/nar/gkr688

Haas, B. J., Dobin, A., Li, B., Stransky, N., Pochet, N., & Regev, A. (2019, Oct 21). Accuracy assessment of fusion transcript detection via read-mapping and de novo fusion transcript assembly-based methods. *Genome Biol, 20*(1), 213. https://doi.org/10.1186/s13059-019-1842-9

Han, C., Sun, L. Y., Wang, W. T., Sun, Y. M., & Chen, Y. Q. (2019, Oct 25). Non-coding RNAs in cancers with chromosomal rearrangements: The signatures, causes, functions and implications. *J Mol Cell Biol, 11*(10), 886–898. https://doi.org/10.1093/jmcb/mjz080

Heyer, E. E., Deveson, I. W., Wooi, D., Selinger, C. I., Lyons, R. J., Hayes, V. M., O'Toole, S. A., Ballinger, M. L., Gill, D., Thomas, D. M., Mercer, T. R., & Blackburn, J. (2019, Mar 27). Diagnosis of fusion genes using targeted RNA sequencing. *Nat Commun, 10*(1), 1388. https://doi.org/10.1038/s41467-019-09374-9

Khordadmehr, M., Shahbazi, R., Ezzati, H., Jigari-Asl, F., Sadreddini, S., & Baradaran, B. (2019, Jun). Key microRNAs in the biology of breast cancer; emerging evidence in the last decade. *J Cell Physiol, 234*(6), 8316–8326. https://doi.org/10.1002/jcp.27716

Lagos-Quintana, M., Rauhut, R., Yalcin, A., Meyer, J., Lendeckel, W., & Tuschl, T. (2002, Apr 30). Identification of tissue-specific microRNAs from mouse. *Curr Biol, 12*(9), 735–739. https://doi.org/10.1016/s0960-9822(02)00809-6

Lau, N. C., Lim, L. P., Weinstein, E. G., & Bartel, D. P. (2001, Oct 26). An abundant class of tiny RNAs with probable regulatory roles in *Caenorhabditis elegans*. *Science, 294*(5543), 858–862. https://doi.org/10.1126/science.1065062

Mackowiak, S. D. (2011, Dec). Identification of novel and known miRNAs in deep-sequencing data with miRDeep2. *Curr Protoc Bioinform, 12*, 10. https://doi.org/10.1002/0471250953.bi1210s36

Mansoori, B., Mohammadi, A., Ghasabi, M., Shirjang, S., Dehghan, R., Montazeri, V., Holmskov, U., Kazemi, T., Duijf, P., Gjerstorff, M., & Baradaran, B. (2019, Jun). miR-142–3p as tumor suppressor miRNA in the regulation of tumorigenicity, invasion and migration of human breast cancer by targeting Bach-1 expression. *J Cell Physiol, 234*(6), 9816–9825. https://doi.org/10.1002/jcp.27670

Role of Transcriptomics in Identifying Fusion Genes and Chimeric RNAs   151

Missiaglia, E., Shepherd, C. J., Aladowicz, E., Olmos, D., Selfe, J., Pierron, G., Delattre, O., Walters, Z., & Shipley, J. (2017, Jan 28). MicroRNA and gene co-expression networks characterize biological and clinical behavior of rhabdomyosarcomas. *Cancer Lett, 385,* 251–260. https://doi.org/10.1016/j.canlet.2016.10.011

Mitelman, F., Johansson, B., & Mertens, F. (2007, Apr). The impact of translocations and gene fusions on cancer causation. *Nat Rev Cancer, 7*(4), 233–245. https://doi.org/10.1038/nrc2091

Okonechnikov, K., Imai-Matsushima, A., Paul, L., Seitz, A., Meyer, T. F., & Garcia-Alcalde, F. (2016). InFusion: Advancing discovery of fusion genes and chimeric transcripts from deep RNA-sequencing data. *PLoS One, 11*(12), e0167417. https://doi.org/10.1371/journal.pone.0167417

Parker, B. C., & Zhang, W. (2013, Nov). Fusion genes in solid tumors: An emerging target for cancer diagnosis and treatment. *Chin J Cancer, 32*(11), 594–603. https://doi.org/10.5732/cjc.013.10178

Ramakrishna, S., & Muddashetty, R. S. (2019, Apr 19). Emerging role of microRNAs in dementia. *J Mol Biol, 431*(9), 1743–1762. https://doi.org/10.1016/j.jmb.2019.01.046

Ransohoff, J. D., Wei, Y., & Khavari, P. A. (2018, Mar). The functions and unique features of long intergenic non-coding RNA. *Nat Rev Mol Cell Biol, 19*(3), 143–157. https://doi.org/10.1038/nrm.2017.104

Rodriguez-Martin, B., Palumbo, E., Marco-Sola, S., Griebel, T., Ribeca, P., Alonso, G., Rastrojo, A., Aguado, B., Guigo, R., & Djebali, S. (2017, Jan 3). ChimPipe: Accurate detection of fusion genes and transcription-induced chimeras from RNA-seq data. *BMC Genomics, 18*(1), 7. https://doi.org/10.1186/s12864-016-3404-9

Taniue, K., & Akimitsu, N. (2021, Feb 4). Fusion genes and RNAs in cancer development. *Noncoding RNA, 7*(1), 12. https://doi.org/10.3390/ncrna7010010

Teixeira, M. R. (2006, Dec). Recurrent fusion oncogenes in carcinomas. *Crit Rev Oncog, 12*(3–4), 257–271. https://doi.org/10.1615/critrevoncog.v12.i3-4.40

Zhang, S. L., & Liu, L. (2015, Feb). microRNA-148a inhibits hepatocellular carcinoma cell invasion by targeting sphingosine-1-phosphate receptor 1. *Exp Ther Med, 9*(2), 579–584. https://doi.org/10.3892/etm.2014.2137

Zhao, J., Chen, Q., Wu, J., Han, P., & Song, X. (2017, Jul 31). GFusion: An effective algorithm to identify fusion genes from cancer RNA-seq data. *Sci Rep, 7*(1), 6880. https://doi.org/10.1038/s41598-017-07070-6

Zhao, W. W., Wu, M., Chen, F., Jiang, S., Su, H., Liang, J., Deng, C., Hu, C., & Yu, S. (2015). Robertsonian translocations: An overview of 872 Robertsonian translocations identified in a diagnostic laboratory in China. *PLoS One, 10*(5), e0122647. https://doi.org/10.1371/journal.pone.0122647

6 MiRNA and RNA-seq

Robert Morris and Feng Cheng
University of South Florida

CONTENTS

6.1 MiRNAs 153
    6.1.1 What Are miRNAs? 153
    6.1.2 Generation of miRNAs... 154
6.2 lncRNAs 156
    6.2.1 What Are lncRNAs?... 156
    6.2.2 Generation and Structure of lncRNAs 156
6.3 miRDeep2... 157
    6.3.1 Methodology of miRDeep2.. 157
    6.3.2 MirDeep2 Galaxy Example...161
6.4 Applications... 165
    6.4.1 Non-coding RNAs in Hypertrophic Cardiomyopathy................ 165
    6.4.2 miRNA-Regulated Drug-Pathway (MRDP) Network................ 166
6.5 Summary 167
Keywords and Phrases 167
Bibliography 168

6.1 MIRNAS

6.1.1 WHAT ARE MIRNAS?

MicroRNAs (miRNAs) are a class of small endogenous regulatory non-coding RNAs (ncRNAs) that function as post-transcriptional suppressors of gene expression in such pathways as proliferation, regulation of the cell cycle, and apoptosis. Long non-coding RNAs (lncRNAs) have some degree of overlap in function as miRNAs in that they are capable of suppressing the translation of target transcripts; however, they have also been implicated in tumorigenesis, remodeling of chromatin structure, and miRNA sponging, the latter of which is defined as the binding of miRNAs to lncRNAs and the subsequent suppression of target miRNA function.

MiRNAs are a class of single-stranded ncRNAs that are approximately 22 bp in length and are mechanistically involved in the post-transcriptional regulation of at least 60% of the human genome. MiRNAs were initially discovered in the model organism *C. elegans* in 1993 as a repressor of larval development by inhibiting the function of the *lin-14* gene and miRNAs

DOI: 10.1201/9781003174028-6

have significantly expanded upon our understanding of gene regulation and the functionality of RNAs in biological systems.

Within the genome, miRNA sequences are frequently positioned within fragile areas of chromosomes where breakage induced by replication stress is most common. Most species of miRNAs originate as double-stranded DNA transcripts transcribed by RNA polymerase II (RNA pol II), although some miRNAs are instead transcribed by RNA polymerase III (RNA pol III). They are either present as isolated polycistronic clusters that code for multiple miRNAs or are located within the intronic regions of other genes.

6.1.2 Generation of miRNAs

Figure 6.1 illustrates the typical hairpin loop secondary structure of a mature miRNA. Initiation of the processing of nascent miRNAs, which are termed pri-miRNA, is mediated by the binding of the dsRNA protein DiGeorge syndrome critical region gene 8 (DGCR8) and the subsequent recruitment of the RNA III enzyme Drosha, ultimately resulting in what is referred to as the Microprocessor. Drosha-mediated cleavage of the pri-miRNA by the Microprocessor involves the excision of the lower stem of the pri-miRNA, producing a pre-miRNA of approximately 60 bp in length which contains a single-stranded 3' overhang.

The two nucleotides (nt) overhang of the pre-miRNA then gets recognized by the nuclear export protein XPO5 and is directly transported to the cytoplasm in a RAN-GTP-dependent manner for further structural processing. After being transported through the nuclear pores to the cytoplasm, the 2-nt 3' overhang of the pre-miRNA is recognized by an additional RNase III enzyme, Dicer, which triggers Dicer-mediated cleavage of the 3' terminal loop and yields the mature miRNA that is capable of transcript targeting (Figure 6.2).

Non-canonical miRNA species have recently been *discovered* in Drosophila and mammals which undergo alternative maturation events to ultimately yield a mature miRNA, the most common of which are termed mirtrons. Bypassing the

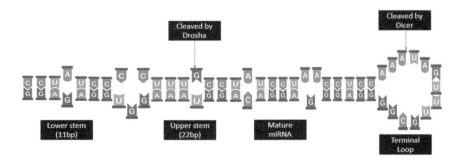

FIGURE 6.1 Overview of the general hairpin loop structure of many miRNA species. Positions in the hairpin loop that are cleaved by the two RNAses Drosha and Dicer are noted.

MiRNA and RNA-seq

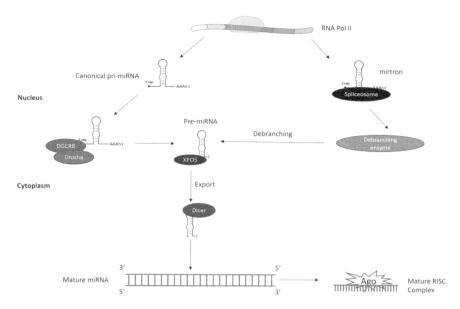

FIGURE 6.2 Processing pathways for both canonical and non-canonical (mirtrons) are depicted.

Drosha-mediated cleavage required in the canonical miRNA processing pathway, mirtrons are extremely small intronic sequences that once spliced out of a mature mRNA will undergo a conformational shift that results in a pre-miRNA-mimicking stem-loop structure. This pre-miRNA can subsequently be transported to the cytoplasm by XPO5 and further processed by Dicer in a manner similar to the canonical processing pathway.

Once the pri-miRNA is fully processed into the mature miRNA, miRNAs are subsequently sorted and loaded into one of several different Argonaute (AGO) proteins in a process dependent on the terminal nucleotides of the miRNA, the structure of the miRNA, thermodynamic stability of the Ago-miRNA interaction, and the location in which the bound AGO protein is subsequently transported. The loading of mature double-stranded miRNAs into one of the AGO proteins is promoted by energy derived from ATP hydrolysis. In addition, the association between the AGO proteins and the chaperone complex comprising heat shock cognate 70 (HSC70) and heat shock protein 90 (HSP90) (18) aids in the promotion of miRNA loading. The association of this HSC70/HSP90 chaperone complex with the AGO protein alters the structural conformation of the AGO protein into one that is capable of accepting a double-stranded mature miRNA. This immature RNA-induced silencing complex (RISC) complex in its 'open' conformation must go through an additional maturation step in which one strand of the associated miRNA, denoted as the passenger strand or miR*, is ejected from the complex. The selection for which strand is ejected is primarily determined by the

156   RNA-seq in Drug Discovery and Development

thermodynamic stability of each strand where the miRNA with the weakest 5'
binding remains bound to the RISC complex as the guide strand, or miR, and the
other strand is cleaved.

6.2 lncRNAs

6.2.1 WHAT ARE LNCRNAS?

In contrast to short ncRNA such as miRNAs, lncRNAs are defined as transcripts
longer than 200 bp and are primarily transcribed by RNA Pol II. The promoter
regions of lncRNAs may lie solely within the intergenic regions of coding genes,
termed long intergenic non-coding RNAs, or may have some degree of overlap
with the sequences of protein-coding genes.

6.2.2 GENERATION AND STRUCTURE OF LNCRNAS

In most cases, lncRNAs are transcribed co-transcriptionally alongside the protein-
coding genes in which they share intergenic or genic sequences with; however,
transcriptome-wide analyses have indicated that lncRNA transcripts are several
orders of magnitude lower in abundance than their corresponding protein-coding
mRNAs. This may be a result of the frequent pausing of RNA Pol II when reading
the lncRNA sequence, thus promoting frequent premature termination and degradation of the lncRNA transcript through nonsense-mediated decay. In contrast,
protein-coding sequences experience significantly less replication fork stalling,
resulting in a greater number of functional mRNA products. Once transcribed,
lncRNA transcripts are capped with a 5' methyl-guanosine and, depending on
the transcript species, will also be polyadenylated on its 3' end. In addition to co-
transcription, lncRNAs also undergo coupled alternative splicing events in which
the removed intronic sequences undergo a subsequent splicing event after they
are spliced out of the spliced mRNA transcript. However, the splicing efficiency
of these nascent pre-lncRNAs is considerably lower than that of pre-mRNA and
may be a result of relatively weaker 3' splice sites in most lncRNAs as well as the
lower recruitment of U2AF, an essential splicing factor.

LncRNAs are categorized based upon the functionality of their transcripts,
their localization patterns, their structural composition, and where the lncRNA
promoter region is located such as intergenically or within the coding region
of another gene. The promoters of most eukaryotic genes are bidirectional in
nature and thus can be transcribed in either a sense or antisense direction. While
mRNAs are almost exclusively transcribed with sense directionality, lncRNAs
can be transcribed in either direction within introns or intergenic regions and
thus can be categorized as sense, antisense, bidirectional, intronic, or intergenic
lncRNAs (Figure 6.3).

The primary structure of lncRNAs is highly variable between different transcript species; however, the secondary and tertiary structures of these same
transcripts are heavily conserved across the different categories of lncRNAs.

MiRNA and RNA-seq

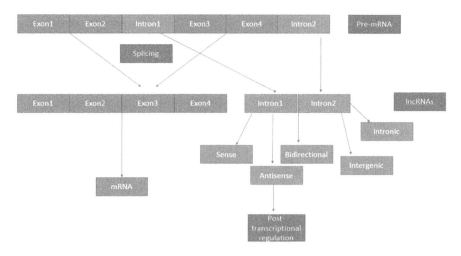

FIGURE 6.3 Overview of the types of lncRNAs and where they are generated from.

LncRNAs primarily function as post-transcriptional regulators through acting as miRNA sponges, binding to target proteins to promote or inhibit function, promoting proper localization of splicing factors, and through serving as adaptors for protein-protein interactions.

6.3 MiRDeep2

6.3.1 Methodology of miRDeep2

MiRDeep2 is one of the most commonly used tools to assess changes in the ncRNA transcriptome, particularly fluctuations in miRNA expression, between conditions, dysregulation of which is closely related to the generation of fusion genes in cancer progression. The miRDeep2 pipeline has the following three distinct modules:

1. **Mapper module**: This module processes the Illumina raw reads input and maps them to a reference genome.
2. **Quantifier module**: This module generates a table of read counts for the mapped ncRNAs produced in the mapper module.
3. **MiRDeep2 module**: Primary tool that identifies known and novel ncRNA species from sequencing data.

Figure 6.4 provides an outline of each step for each of the miRDeep2 modules. The mapper module requires a FASTQ file containing our raw reads, a .seq file containing the sequence of our 3' adaptor, and a reference genome in FASTA

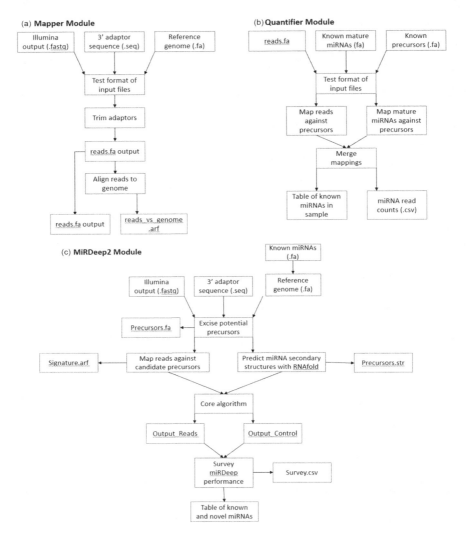

FIGURE 6.4 Steps utilized by the mapper module (a), the quantifier module (b), and the miRDeep2 module (c) are presented. Individual steps are outlined in red and output files generated by each module are outlined in blue.

format as input. The miRDeep2 mapper module maps the raw reads to the reference genome in the following steps:

1. Tests for appropriate format of input files.
2. Clips 3' adaptor sequences and collapses reads. In order to accomplish this, each read sequence is scanned for any matches to the first six nucleotides of the provided 3' adaptor sequence beginning at position 18 in

MiRNA and RNA-seq 159

the read. If a six-nucleotide match is not detected, then the module will subsequently scan the read for a five-nucleotide match with the adaptor sequence. This process continues until a match to the adaptor sequence is detected, at which point the matched sequence and every subsequent nucleotide in the read are clipped.

3. Repeats step 2 for all reads. Reads with no sequence that matches the' 3' adaptor sequence are not clipped, but are retained in the pool of transcripts.

4. All reads with identical sequences are collapsed into a single read and a note is made in the FASTA file as to the count for each sequence.

5. Maps the remaining processed reads to the reference genome using Bowtie. By default, only alignments with zero mismatches in the seed sequence are mapped, that is no mismatches are allowed in the first 18 nucleotides of the read sequence. However, you may modify the number of allowed mismatches in the sequence beyond the first 18 nucleotides (default=two mismatches).

6. The list of remaining processed reads and the alignments are outputted in .fa and .arf format, respectively.

Next, the reads in .fa format generated from the mapper module in the previous step are then used as input for the quantifier module along with downloaded lists of known mature miRNAs and known miRNA precursors in .fa format. Like reference genomes, these files are species specific and can be downloaded from a public database such as miRbase. Further instruction on how to access these files can be found in the subsequent example for this pipeline. The quantification of miRNA transcripts from the sample by the quantifier module occurs in the following order:

1. Like the mapper module, the quantifier module tests for the appropriate format for the input files.

2. Maps the sequenced reads and the known mature miRNAs against the known list of precursor miRNAs. Bowtie is used to build a Burrows–Wheeler transform index of the known miRNA precursors. When mapping the sequenced reads, one mismatch is allowed by default for output reports; however, no mismatches are allowed for a given sequence to be reported when mapping mature miRNA sequences. This is because annotated precursor miRNAs should contain the exact sequence as the mature miRNA.

3. The two read maps are then intersected. A read is assumed to be a true mature miRNA if it falls on the same location on the precursor as well as two nucleotides upstream and five nucleotides downstream of the position. Read sequences that match well to more than one mature miRNA contribute to the counts for each mature miRNA that the reads successfully matched to.

The final module of the miRDeep2 pipeline is next used to identify both known and novel miRNA species. As input, this miRDeep2 module requires a reference genome in FASTA format, the reads .fa file generated from the mapper module, the reads_vs_genome.arf file produced from the mapper module, and a list of known miRNAs in FASTA format. As output, the miRDeep2 module generates a table that lists both known and novel miRNAs in the provided sequencing data. This is performed by the miRDeep2 module using the following protocol steps:

1. Tests for appropriate format of input files.
2. Potential candidates for miRNA precursors are removed from the reference genome by using the mapped sequenced reads. Mappings of the sequenced reads are first filtered so that only mappings with no mismatches of at least a length of 18 nucleotides are retained. In addition, any mapped reads that are mapped to the genome >5 times are removed. Each strand of the contig is then scanned separately with 5' directionality. In an iterative process, excision occurs when a stack of reads (# of reads≥ 1) is found. However, if a stack with a greater number of reads is found within 70 nucleotides downstream of the first stack, the new stack with the greater number of reads is instead excised. The sequence encompassed by the highest read stack is then excised twice, one that includes 70 nucleotides upstream and 20 nucleotides downstream and another that has the matched sequence flanked by 20 nucleotides upstream and 70 nucleotides downstream. If the number of candidate precursors is less than 50,000, the list of candidates will be reported in an output file. However, in instances where the number of candidates exceeds 50,000, the excision process is repeated with the minimum number of reads required in a stack increased by 1. This process continues until the number of candidate miRNA precursors ≤50,000.
3. The signature.arf file is made by using the Bowtie-build tool to construct a Burrows–Wheeler transform index of the precursor miRNAs. ARF stands for Asset Reporting Format.
4. The sequencing reads as well as the known miRNAs are individually mapped to the BWT index created in step 3.
5. The two mapping files generated from step 4 are then concatenated and sorted by precursor ID.
6. RNA secondary structures are predicted using the RNAfold tool and p-values are optionally calculated.
7. The miRDeep2 core algorithm then assigns scores to each candidate miRNA precursor with all precursors assigned a score of ≥50 being retained. This is repeated 100 times using the same parameters to generate control runs.
8. The distribution of scores between the initial run and the 100 control reads as well as performance statistics are calculated using a scale of −10 to 10.
9. The number of known miRNAs recovered is assumed to be approximately equal to the number of known mature miRNA transcripts that mapped without any mismatches to at least one excised candidate

MiRNA and RNA-seq

precursor miRNA. In addition, the number of recovered known miR-NAs is estimated as the number of known mature miRNAs that mapped to at least one hairpin structure that exceeded the set score cutoff. The fraction of true miRNAs is estimated using the formula (Eq. 6.1):

$$t = \frac{\left(\text{novel miRNAs} - \text{estimated false positive miRNAs}\right)}{\text{novel miRNAs}} \tag{6.1}$$

6.3.2 MirDeep2 Galaxy Example

For the following example, we will utilize a public dataset of a small RNA library derived from exosomes (extracellular vesicles) secreted by human natural killer cells treated with Il-15 (GEO accession: GSE150342). Due to compatibility issues with the Galaxy server, the miRDeep2 module, which primarily is used to identify novel miRNA transcripts, will not be used. Instead, we will use the mapper and quantifier modules to identify miRNAs that are present in the sample and create a count table for those identified in our sample. The following parameters (Figure 6.5) were used for the mapper module to generate the collapsed reads and the ARF-formatted mapping outputs:

1. Use the downloaded raw reads (FASTA or fastq format) as input for deep sequencing reads.
2. Set remove reads with non-standard reads to yes.
3. Set convert RNA to DNA alphabet to no.
4. Set clip 3' adaptor sequence to clip sequence.
5. Set the sequence to clip option to TCACTTCGTATGCCGTCTTCTGC TTG. Information of the adaptor sequence can be found in the sample background for each entry in the GEO database.
6. For reference genome, set it to Human (Homo sapiens): hg38.
7. Leave the remaining parameters to their default options.

Figure 6.6 shows excerpts from the collapsed reads (1) and the mapped reads in ARF format (2) that are generated from the mapper module. The ARF format is a modified version of the FASTA format with 13 columns and is summarized in Table 6.1. Before we use these outputs as input for the quantifier module, we first need to download two separate reference files: one of all known miRNA precursors and one for all known mature miRNAs. These can be downloaded from multiple public databases, but the steps below will detail how to access the sequences from the miRbase database:

1. Go to https://mirbase.org/ and click the browse tab near the top of the screen.
2. Select the organism of interest, e.g., humans in this example.

162 RNA-seq in Drug Discovery and Development

FIGURE 6.5 Parameters used to generate mapper module output files.

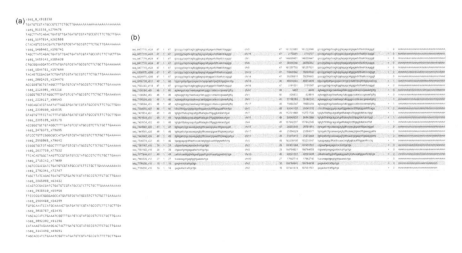

FIGURE 6.6 Excerpts from the collapsed reads in FASTA format (a) and the ARF-formatted mapped reads (b) are presented.

MiRNA and RNA-seq

TABLE 6.1
Overview of the General Structure for ARF-Formatted Files

Column #	Description
1	Read identifier
2	Length of the read sequence
3	Start position in mapped read sequence
4	End position in mapped read sequence
5	Read sequence
6	Identifier of the genome part to which a read is mapped to. This is either a scaffold id or a chromosome name
7	Length of the genome sequence that a read is mapped to
8	Start position in the genome where a read is mapped to
9	End position in the genome where a read is mapped to
10	Genome sequence to which a read is mapped
11	Genome strand information. Plus means the read is aligned to the sense strand of the genome. Minus means it is aligned to the antisense strand of the genome
12	Number of mismatches in the read mapping
13	Edit string in which correct matches are indicated by lowercase 'm' and mismatches are denoted by uppercase 'M'

3. On the next page, scroll down to the bottom of the page and set *select sequence type* to stem-loop sequence. This corresponds to the known precursor sequences.
4. Right below that option, set *choose output format* to unaligned fasta format.
5. Click select all to highlight every miRNA sequence and then click *fetch sequences*.
6. Copy and paste all of the sequences into the notepad app on your computer and save as a .txt file.
7. Return to the previous page and set select sequence type to mature sequence.
8. Repeat steps 5 and 6 to retrieve the reference sequences for mature miRNAs and save in the notepad app as a .txt file.
9. Upload the two .txt files into your Galaxy history.

Now that we have the two output files from the mapper module as well as our reference sequences, we can now generate count tables for the detected miRNA sequences in our sample, found in our collapsed reads file, using the quantifier tool. The following parameters were used to generate the output count table and

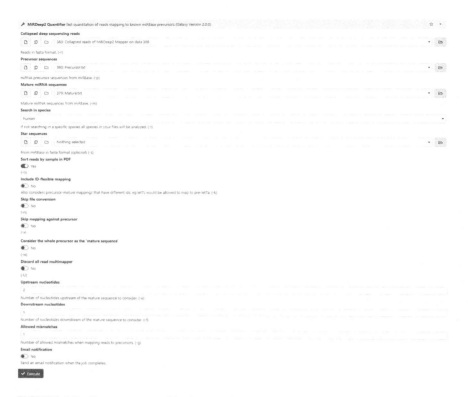

FIGURE 6.7 Parameters used in the quantifier module.

html formatted list of known miRNAs in our sample (Figure 6.7) with sample excerpts of the outputs presented in Figure 6.8:

1. For *collapse deep sequencing reads*, use the collapsed reads file generated by the mapper module as input.
2. For *precursor sequences*, use the precursor.txt file containing the precursor sequences from miRbase.
3. For *mature sequences*, use the mature.txt file containing the mature miRNA sequences from miRbase.
4. Set *search in species* option to human.
5. Leave the remaining options to their default values and click execute.

As you can see from the output, the precursor with the highest expression count was miR-339 with ten counts. In instances where more than one mature miRNA is derived from the same precursor as is the case with miR-339-3p and miR-339-5p, both the output table and graphical representation will depict how many of each mature miRNA transcript was expressed. In this example, all ten counts for the miR-339 precursor corresponded to the miR-339-5p miRNA species.

MiRNA and RNA-seq

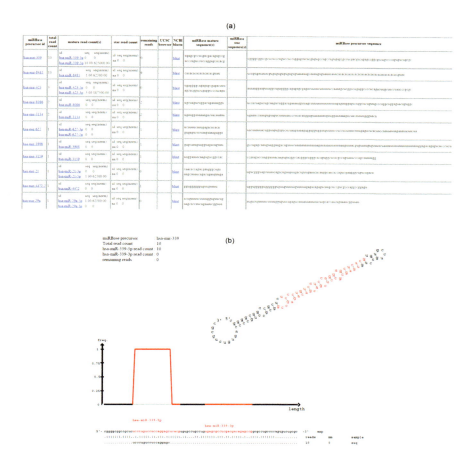

FIGURE 6.8 Tabular output for top expressed miRNAs (a) and predicted structure for miR-339-5p (b) are presented.

6.4 APPLICATIONS

In this section, we will introduce two applications in which RNA-seq technology is used to identify differential ncRNA profiles in two disease states.

6.4.1 NON-CODING RNAs IN HYPERTROPHIC CARDIOMYOPATHY

First, a study by Li *et al.* sought to identify differentially expressed genes (DEGs) and ncRNA species between healthy patients and patients presenting with hypertrophic cardiomyopathy, a condition characterized by abnormally thick heart muscles. Public datasets derived from the GEO data repository including GSE36961 (mRNA), GSE36946 (miRNA), GSE68316 (lncRNA), GSE68316 (mRNA), and GSE32453 (mRNA) were analyzed in this study. Limma was used to identify

DEGs, differentially expressed miRNAs (DEmiRs), and differentially expressed lncRNAs (DElncs) and the online tool GEO2R was used to compare the two GSE68316 datasets to identify differentially expressed RNA species using a t-test statistical analysis. Prediction of DElncs was carried out using the three public databases DIANA-LncBase, starBase, and LNCipedia. Any overlapping DElncs identified from both the Limma analysis and the database prediction based on sequence complementarity binding were retained for further downstream analysis. A similar procedure was used to identify putative miRNA-mRNA pairs by using the public databases TargetScan, miRTarBase, and miRbase 21. Any miRNAs identified from both the Limma analysis for differential expression and the prediction based on miRNA-mRNA pairs were saved for further analysis. In addition, any putative DElncs and DEGs that shared miRNA binding loci were considered one possible lncRNA–miRNA–mRNA interaction.

Pathway analysis and functional annotation were performed using The Database for Annotation, Visualization and Integrated Discovery (DAVID) and Kyoto Encyclopedia of Genes and Genomes/Gene Ontology (KEGG/GO), respectively, and protein-protein interaction networks were constructed using Cytoscape to elucidate lncRNA–miRNA–mRNA interactions. In total, 2234 DElncs (1120 upregulated and 1114 downregulated), 5 DEmiRs (2 upregulated and 3 downregulated), and 42 DEGs (35 upregulated and 7 downregulated) were identified across all of the aforementioned public Gene Expression Omnibus (GEO) datasets. Based on the gene ontology analysis, the identified DEGs were primarily linked to stress fiber formation and calcium ion binding while KEGG pathway analysis indicated that the hypoxia inducible factor-1, transforming growth factor-β, and tumor necrosis factor signaling pathways were the predominant pathways. Cytoscape predicted 1,086 total lncRNA–miRNA–mRNA interactions involving 67 lncRNAs, 5 miRNAs, and 25 mRNAs that were mined in previous steps. Finally, drug prediction was performed using DrugBank in order to propose new drug treatments and identify possible biomarker targets in hypertrophic cardiomyopathy patients, many of which included various angiotensin-converting enzyme inhibitors or beta-blockers.

6.4.2 miRNA-Regulated Drug-Pathway (MRDP) Network

In the second study, *Cao et al.* sought to construct an MRDP network in order to repurpose existing therapeutics in order to find novel treatments for myasthenia gravis (MG), an autoimmune neuromuscular disorder characterized by rapid onset of muscle fatigue. MG risk genes were identified by combing through 9,474 manuscripts written prior to 2015 and stored in the PubMed database. MG risk genes were identified based on the following criteria:

1. The gene in question was present in at least five MG samples including blood samples and thymic tissue samples.
2. The gene was detected using scientifically valid methods including microarrays and reverse transcription-polymerase chain reaction RT-PCR.

MiRNA and RNA-seq

3. The gene was significantly differentially expressed between healthy samples and MG samples.

In addition, MG risk genes were collected from up-to-date databases including DisGeNET, Functional Disease Ontology Annotation (FunDO), and the Genetic Association Database. MG risk miRNAs were manually collected from the public databases Human microRNA Disease Database and PhenomiR as well as from a PubMed literature search using the search keywords 'miRNA'/'microRNA'/'miR' paired with 'myasthenia gravis'. Target genes of these identified MG risk miRNAs were subsequently collected from five experimentally validated miRNA databases including miRTarBase, miRecord, miRSel, miRWalk, and miR2Disease.

Overall, 162 MG risk genes, of which 123 risk genes were identified through the PubMed literature search and 39 were mined from public databases, as well as 85 miRNAs were identified. Drugs and their target genes were derived from DrugBank and pathways implicated in MG pathology were identified using the list of collected DEGs. Forty-five distinct MG risk pathways were identified that achieved statistical significance including cytokine–cytokine receptor interactions and intestinal immune network for IgA. miRNA-146a was found to be a critical regulator of many of the identified pathways implicated in MG pathology. All of this collected information was then combined to generate an MRDP that illustrated interactions between miRNAs and drugs that synergistically regulated pathways related to MG pathology and progression.

In total, 25 drugs including rituximab, adalimumab, sunitinib, and muromonab were found to be possible novel treatments for MG. As you can see, RNA-seq data can be used to construct interaction networks between target mRNAs, miRNAs, and drugs for the purposes of repurposing existing drugs for new therapeutic treatments as well as identify differentially expressed ncRNA species that may contribute to the progression of a particular disease pathology.

6.5 SUMMARY

In this chapter, we introduced fusion genes and the processes that may occur to generate novel fusion products. In addition, we introduced ncRNAs, specifically miRNAs and lncRNAs, as well as the molecular mechanism in which miRNAs are transcribed. Finally, we introduced the three modules of a tool, miRDeep2, that can detect differential expression of ncRNAs, particularly miRNAs, both within and across samples.

KEYWORDS AND PHRASES

After reading this chapter, you should be able to demonstrate familiarity with the following words and phrases:

- Be able to describe the mechanisms involved in the transcription of ncRNAs.

168 RNA-seq in Drug Discovery and Development

- Be able to differentiate between miRNAs and lncRNAs.
- Understand the three modules of miRDeep2 and be able to successfully run it in Galaxy.
- Be able to understand and interpret miRDeep2 output files such as structure files and the list of most identified mature sequences.

BIBLIOGRAPHY

Bartonicek, N., Maag, J. L., & Dinger, M. E. (2016, May 27). Long noncoding RNAs in cancer: Mechanisms of action and technological advancements. *Mol Cancer*, 15(1), 43. https://doi.org/10.1186/s12943-016-0530-6

Cao, Y., Lu, X., Wang, J., Zhang, H., Liu, Z., Xu, S., Wang, T., Ning, S., Xiao, B., & Wang, L. (2017, Feb). Construction of an miRNA-regulated drug-pathway network reveals drug repurposing candidates for myasthenia gravis. *Int J Mol Med*, 39(2), 268–278. https://doi.org/10.3892/ijmm.2017.2853

Czech, B., & Hannon, G. J. (2011, Jan). Small RNA sorting: Matchmaking for Argonautes. *Nat Rev Genet*, 12(1), 1920–1331. https://doi.org/10.1038/nrg2916

Dhanoa, J. K., Sethi, R. S., Verma, R., Arora, J. S., & Mukhopadhyay, C. S. (2018). Long non-coding RNA: Its evolutionary relics and biological implications in mammals. *J Anim Sci Technol*, 60, 25. https://doi.org/10.1186/s40781-018-0183-7

Ding, X., Zhu, L., Ji, T., Zhang, X., Wang, F., Gan, S., Zhao, M., & Yang, H. (2014). Long intergenic non-coding RNAs (LincRNAs) identified by RNA-seq in breast cancer. *PLoS One*, 9(8), e103270. https://doi.org/10.1371/journal.pone.0103270

Friedlander, M. R., Mackowiak, S. D., Li, N., Chen, W., & Rajewsky, N. (2012, Jan). miRDeep2 accurately identifies known and hundreds of novel microRNA genes in seven animal clades. *Nucleic Acids Res*, 40(1), 3720–1352. https://doi.org/10.1093/nar/gkr688

Han, C., Sun, L. Y., Wang, W. T., Sun, Y. M., & Chen, Y. Q. (2019, Oct 25). Non-coding RNAs in cancers with chromosomal rearrangements: The signatures, causes, functions and implications. *J Mol Cell Biol*, 11(10), 88620–13898. https://doi.org/10.1093/jmcb/mjz080

Khordadmehr, M., Shahbazi, R., Ezzati, H., Jigari-Asl, F., Sadreddini, S., & Baradaran, B. (2019, Jun). Key microRNAs in the biology of breast cancer; emerging evidence in the last decade. *J Cell Physiol*, 234(6), 831620–138326. https://doi.org/10.1002/jcp.27716

Lagos-Quintana, M., Rauhut, R., Yalcin, A., Meyer, J., Lendeckel, W., & Tuschl, T. (2002, Apr 30). Identification of tissue-specific microRNAs from mouse. *Curr Biol*, 12(9), 73520–13739. https://doi.org/10.1016/s0960-9822(02)00809-6

Lau, N. C., Lim, L. P., Weinstein, E. G., & Bartel, D. P. (2001, Oct 26). An abundant class of tiny RNAs with probable regulatory roles in Caenorhabditis elegans. *Science*, 294(5543), 85820–13862. https://doi.org/10.1126/science.1065062

Li, J., Wu, Z., Zheng, D., Sun, Y., Wang, S., & Yan, Y. (2019, Jul). Bioinformatics analysis of the regulatory lncRNAmiRNAmRNA network and drug prediction in patients with hypertrophic cardiomyopathy. *Mol Med Rep*, 20(1), 54920–13558. https://doi.org/10.3892/mmr.2019.10289

Mackowiak, S. D. (2011, Dec). Identification of novel and known miRNAs in deep-sequencing data with miRDeep2. Curr Protoc Bioinform, 12, 14. https://doi.org/10.1002/0471250953.bi1210s36

Mansoori, B., Mohammadi, A., Ghasabi, M., Shirjang, S., Dehghan, R., Montazeri, V., Holmskov, U., Kazemi, T., Duijf, P., Gjerstorff, M., & Baradaran, B. (2019, Jun). miR-142-3p as tumor suppressor miRNA in the regulation of tumorigenicity, invasion and migration of human breast cancer by targeting Bach-1 expression. *J Cell Physiol*, 234(6), 981620–139825. https://doi.org/10.1002/jcp.27670

Missiaglia, E., Shepherd, C. J., Aladowicz, E., Olmos, D., Selfe, J., Pierron, G., Delattre, O., Walters, Z., & Shipley, J. (2017, Jan 28). MicroRNA and gene co-expression networks characterize biological and clinical behavior of rhabdomyosarcomas. *Cancer Lett*, 385, 25120–13260. https://doi.org/10.1016/j.canlet.2016.10.011

Ramakrishna, S., & Muddashetty, R. S. (2019, Apr 19). Emerging role of microRNAs in dementia. *J Mol Biol*, 431(9), 174320–131762. https://doi.org/10.1016/j.jmb.2019.01.046

Ransohoff, J. D., Wei, Y., & Khavari, P. A. (2018, Mar). The functions and unique features of long intergenic non-coding RNA. *Nat Rev Mol Cell Biol*, 19(3), 14320–13157. https://doi.org/10.1038/nrm.2017.104

Zhang, S. L., & Liu, L. (2015, Feb). microRNA-148a inhibits hepatocellular carcinoma cell invasion by targeting sphingosine-1-phosphate receptor 1. *Exp Ther Med*, 9(2), 57920–13584. https://doi.org/10.3892/etm.2014.2137

7 Toxicogenomics and RNA-seq

Robert Morris, Kyle Eckhoff,
Rebecca Polsky, and Feng Cheng
University of South Florida

CONTENTS

7.1	Introduction of Toxicology	172
	7.1.1 What Is Toxicology?	172
	7.1.2 Mechanisms of Toxicity	172
	7.1.3 *In Vivo* Animal Model in Toxicology	173
7.2	Toxicogenomics	175
	7.2.1 What Is Toxicogenomics?	175
	7.2.2 The Advantage of Toxicogenomics	175
	7.2.3 Limitations of Toxicogenomics:	176
7.3	Methods for Toxicogenomics Data Analysis	176
	7.3.1 Identification of Differentially Expressed Genes	176
	7.3.2 Gene Networks	178
	7.3.3 Co-Expression Networks	182
	7.3.3.1 Context Likelihood of Relatedness	184
	7.3.3.2 Weighted Gene Co-expression Network Analysis	185
	7.3.4 Signature Matching	187
7.4	Toxicogenomics Databases	188
	7.4.1 Comparative Toxicogenomics Database	188
	7.4.2 Japanese Toxicogenomics Project	196
	7.4.3 DrugMatrix	198
7.5	Comparing Microarray vs. RNA-seq	200
7.6	Summary	202
	Keywords and Phrases	202
	Bibliography	203

DOI: 10.1201/9781003174028-7

171

172 RNA-seq in Drug Discovery and Development

7.1 INTRODUCTION OF TOXICOLOGY

7.1.1 WHAT IS TOXICOLOGY?

The realm of toxicology is defined by the identification of side effects, both advantageous and pejorative, and biological responses upon exposure to physical and chemical agents. In addition, toxicology is concerned with designing increasingly more sensitive detection methods and treatment programs. In order to determine whether or not a toxic effect has occurred in response to a particular exposure, clear endpoints must be established in order to measure the subsequent effects upon exposure. Most toxicological studies have been concerned with identifying observable phenotypic changes in an organism of interest such as skin rashes resulting from irritation and allergic reactions, development of chronic conditions such as cancer or liver disease, or in the most severe cases, death.

Until 1938, very little legislation existed to protect the general population from dangerous chemicals and the legal framework that did exist only allowed for reactionary measures in response to criminal conduct. At the time, there was no systematic foundation for product analysis, safety testing and guidelines, and efficacy proof for drugs and other products that were marketed. The event that precipitated this shift in toxicology and government regulations was the mislabeling of a solution of sulfanilamide and ethylene glycol by the Massengil Company, a mistake that would lead to at least 73 deaths and result in the passing of the Food, Drug, and Cosmetic Act in 1938. Subsequent amendments to the act and a bolstering of the federal regulations in the 1950s and 1960s have created a complex system of required research needed prior to the production and distribution of a chemical or biological product.

7.1.2 MECHANISMS OF TOXICITY

Mechanisms of toxicity can be broadly classified into the following five groups:

1. **On-target or mechanism-based**: The toxicity response results as a byproduct of the desired interaction between the drug compound and its target. For instance, the use of statins to reduce cholesterol by targeting 3-hydroxy-3-methylglutaryl CoA (HMG CoA) in the liver may also impair proper physiological function of posttranslational protein modifications in muscle tissues through the same inhibition of HMG CoA activity.
2. **Hypersensitivity**: Severe allergic reactions that arise from the production of antibodies due to the reaction of drug metabolites with proteins.
3. **Off-target toxicity**: Toxicity results from the binding of drug metabolites to additional targets beyond the targets that the drug was specifically designed to interact with, resulting in undesired adverse events.
4. **Bioactivation**: Metabolites produced by the breakdown of prodrugs adversely modify target proteins and result in toxicity responses.

Toxicogenomics and RNA-seq

Many mechanisms classified in this category are not well elucidated by existing models.

5. **Idiosyncratic**: Very rare toxicity responses (low incidence) that do not clearly adhere to one of the other classification modalities.

7.1.3 *In Vivo* Animal Model in Toxicology

The cornerstone of toxicology is the use of *in vivo* animal models to assess the toxicity of a substance to a biological system. In these whole-animal studies, the research question may be concerned with the subsequent response to an acute exposure or may focus on the effects of long-term exposure to smaller doses of the toxin in question. When assessing a particular drug or compound for toxicity, there are four major components of drug dynamics, termed pharmacokinetics, that are analyzed for the purpose of determining efficacy and safety. In pharmacokinetics, these principles include

1. **Absorption**: A determinant of bioavailability, absorption is defined by the route of entry for a drug or chemical into the biological system. In most cases, a drug first needs to enter the bloodstream before being transported to its site of action. Drugs and chemical compounds typically have higher absorbance when administered in some areas (e.g., oral and intravenous) than others and thus the site of exposure influences the degree of uptake and possible toxicity.
2. **Distribution**: Once in the bloodstream, the drug is allocated unevenly to different organs and tissues. Factors that influence the degree of uptake of a particular drug include the size of the molecule, molecule polarity, localized blood flow rates, and natural barriers such as the blood–brain barrier.
3. **Metabolism**: The breakdown of biological molecules into metabolites occurs primarily in the liver through the use of a large family of redox enzymes known as cytochrome p450. Metabolites may either be more toxic, less toxic, or inert when compared to the initial compound that first entered the body.
4. **Excretion**: To prevent the accumulation of compounds in the system, ingested compounds and metabolites need to be expelled from the body. The two primary methods of excretion include expulsion through urine, which is carried out by the kidneys, and fecal waste, which is carried out by the liver and the gut. Water-soluble molecules will typically be excreted via the urine, while fat-soluble molecules will typically be processed in the liver, passed to the gut, and excreted along with other waste products in the feces.

Figure 7.1 describes the flowchart of toxicology studies. The first step is identification of the exposure of interest. This is succeeded by determining the absorption, distribution, metabolism, and excretion properties of the molecule of interest.

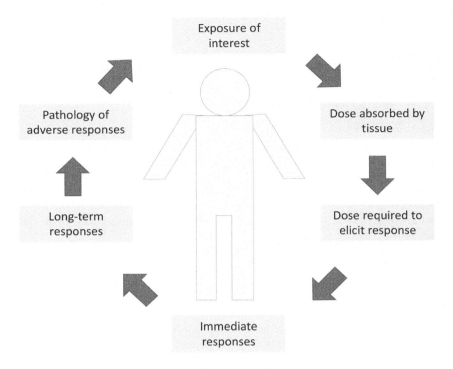

FIGURE 7.1 The progression of a traditional toxicological study.

Next, short- and long-term effects of exposure are monitored and identified. Finally, resulting diseases and their pathology are analyzed and explored. Diseases that ultimately manifest should be able to be linked to the original exposure of interest.

Although in vivo animal models are still the primary means in which toxicology studies are conducted, ethical concerns and a shift in the conscience of the scientific community have prompted a desire to limit animal testing. In addition, *in vivo* animal models have several limitations and disadvantages associated with their use. For example, animal models are extremely time-consuming and expensive to maintain, resulting in further delays in drug approval. In addition, due to physiological differences between humans and laboratory animals used in testing, some drugs may be deemed safe, but are later shown to be toxic in humans during clinical trials. The reverse situation may also occur in which a drug is labeled as unsafe in animal trials yet is actually not toxic at an effective dose in humans. The doses administered to laboratory animals during testing are, in many cases, several folds higher than a normal dose that a human would be exposed to. Thus, in addition to their high costs, the applicability of animal models to drug efficacy and safety in humans may be limited, and drugs that would otherwise be safe for human use in disease management may be lost.

7.2 TOXICOGENOMICS

7.2.1 WHAT IS TOXICOGENOMICS?

Toxicogenomics is defined as the use of analytical techniques and high-throughput technologies to assess alterations in gene expression and protein concentration within cells when exposed to toxic, or potentially toxic, compounds. New advances in 'omics' technology and storage capabilities have precipitated a change in focus toward the utilization of molecular endpoints as opposed to simply observable phenotypic changes. Toxicogenomics measures and compares all gene expression changes among biological samples after toxicant exposure. Profiling with RNA-seq to measure all mRNA transcripts, or by global separation and identification of proteins, has led to the discovery of better descriptors of toxicity, toxicant classification, and exposure monitoring than current indicators. This allows for a greater understanding of the underlying mechanistic nuances that drive toxicity and resulting physiological response.

Toxicogenomics has a variety of additional useful applications. For example, through the use of *in vitro* cell experiments, the short-term and long-term effects of environmental toxin exposure can be assessed in specific cell types. Other applications of toxicogenomics include the analysis of dose–response relationships, extrapolation of results across species, determining variations that influence susceptibility to a particular toxic effect, and hazard screenings. The wide range of functional uses for toxicogenomics provides ample insight into the underlying mechanisms that are induced upon toxin exposure and enables greater ability to design new pharmaceuticals and prevention programs to minimize exposure.

7.2.2 THE ADVANTAGE OF TOXICOGENOMICS

Relative to phenotype-based methods of toxicology, toxicogenomics confers two primary advantages when compared with phenotype-based methods of toxicological studies. While phenotype-based toxicology studies observe overt phenotypic responses and adverse effects due to drug or substance exposure, toxicogenomics grant the ability to discern the underlying mechanisms that confer toxicity-induced effects at the molecular level. High-throughput sequencing techniques such as RNA sequencing or microarray are a sensitive means to assess biological changes due to compound exposure. Expression profiling of transcripts that undergo significant alterations in quantity as a result of drug exposure may provide insight into the mechanistic response that confers toxicity to a given agent or drug.

In toxicogenomics, evaluation of the mechanistic underpinnings of a toxicity response is predominantly carried out by investigating the differentially expressed genes or proteins caused by compounds. These molecules are defined as any genes or proteins that have a statistically significant difference in expression between two groups, usually a control (untreated) group and an experimental (drug-exposed group). Once differentially expressed genes or proteins have been

176 RNA-seq in Drug Discovery and Development

identified between the control group and the exposed group, pathway analysis is then performed in order to identify various biological functions that are modified as a consequence of drug or compound exposure.

In addition, RNA sequencing may be able to identify subtle changes in organ health or function that would otherwise go undetected in less sensitive histopathology analyses. Multiple studies have shown the effectiveness of transcriptomics in the identification of biomarkers of hepatotoxicity and pulmonary toxicity as well as the development of models to predict the likelihood of a toxicity event occurring.

7.2.3 Limitations of Toxicogenomics:

Although toxicogenomics provides unprecedented depth into the workings of toxicity, the complexity of the data output and the interplay of various biological mechanisms can make it difficult to discern the actual underlying causes of toxicity. Differential gene expression (DGE) profiling studies are one of the primary methods to elucidate biological mechanisms of toxicity. However, many drugs do not directly impact the expression of genes but rather directly bind and alter the activity of existing proteins. In addition, the alteration of the activity of a single protein may impact multiple downstream pathways, making it difficult to identify the specific mechanisms that confer a toxicity response. Finally, many confounding variables influence the actions of compounds and subsequent DGE analysis including duration of exposure to the toxin, dose, hormonal status, diet, age, and gender. Thus, when analyzing DGE datasets, multiple factors must be assessed in a system-level context in order to correlate a toxic response with a particular physiological mechanism.

7.3 METHODS FOR TOXICOGENOMICS DATA ANALYSIS

The goal of many toxicogenomic data studies is similar to those discussed in Chapter 3 for general RNA-seq experiments. In many instances, differentially expressed genes are first identified in samples treated with a compound of interest and are then mapped using the Database for Annotation, Visualization and Integrated Discovery (DAVID) or other bioinformatic tools in order to determine enriched pathways that are dysregulated in response to compound exposure. In addition, co-expression networks can be generated to identify genes that elicit a coordinated expression pattern across different samples or timepoints. Finally, these transcriptomic readouts can be used in predictive studies to assess possible mechanistic responses due to side effects for unknown test compounds based upon similarities in differentially expressed genes or enriched pathways.

7.3.1 Identification of Differentially Expressed Genes

Recall that a gene is differentially expressed if its difference in measured expression in two experimental conditions achieves statistical significance. As mentioned

Toxicogenomics and RNA-seq

in Chapter 3, many different tools are available to assess differential expression, each of which makes different statistical assumptions about the underlying probability distribution of the data.

In toxicogenomic studies, a commonly used tool to identify differentially expressed genes is the *R* package *limma*. The *limma* package utilizes linear models to estimate variability both within and between samples. As input, *limma* accepts a matrix of raw read counts where rows represent genomic features (genes, exons, etc.) and columns denote unique samples. Linear modeling is performed for each row with regression coefficients and standard errors estimating each comparison of interest being calculated. In addition, test statistics and *p*-values are calculated. Once each linear model is fitted using the *lmfit* function, whether by the least-squares method or another fitting procedure, the *makeContrasts* function generates a contrast matrix to compute log-fold changes and accompanying *t*-statistics for all comparisons of interest. This estimation of the mean-variance relationship can then be modeled using either a precision weights approach, which utilizes the *voom* function to account for unequal variances across samples due to differential sample size or heteroskedasticity, or by using an empirical Bayes approach.

In Galaxy, we can use the same *featurecounts* files from the atorvastatin experiment utilized in Chapter 3 to run *limma*. The following parameters were utilized to generate the sample *limma* output presented in Figure 7.2:

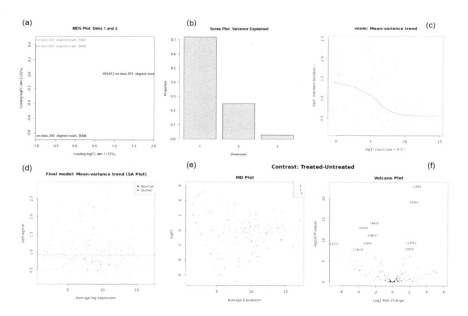

FIGURE 7.2 *Limma* output including MDS plot (a), scree plot that provides further visualization of the variance explained in the MDS plot (b), the mean-variance trend graph generated by the *voom* function (c), the finalized mean-variance Spatial Analysis plot (SA plot) (d), the mean-difference plot (e), and the volcano plot (f).

178 RNA-seq in Drug Discovery and Development

1. Set *differential expression method* option to limma-voom.
2. Set *apply voom with sample quality weights?* option to no.
3. Using separate count files, label *factor 1* as Atorvastatin treatment.
4. Label *group 1* as treated and select the treated *featurecounts* files.
5. Label *group 2* as untreated and select the untreated *featurecounts* files.
6. Label the *contrast of interest* option as Treated-Untreated.
7. Leave the remaining options to their respective default settings.

7.3.2 Gene Networks

Another useful tool for visualization of toxicity mechanisms is a series of figures known as biological interaction networks. Biological interaction networks, of which the most common variant is referred to as a protein–protein interaction (PPI) network, may either be directed or undirected depending on whether or not the edges connecting each node have a particular meaning.

Like previously mentioned interaction networks, each node in a PPI represents one distinct protein, while each edge connecting nodes represents some type of interaction such as activation or inhibition. In directed PPIs, we know the directionality of the flow of information from one node to another and will usually depict signaling networks that display sequentially activated proteins in a signaling cascade. In contrast, the information contained in the edges joining adjacent nodes in undirected PPIs are either unknown or have no biological meaning, the latter of which is the case when showing proteins forming a complex together as opposed to signaling mechanisms. In particular, directed biological networks are much more informative and allow us to observe the downstream cellular response induced by compound exposure.

A variety of different platforms are available in order to construct PPIs, the choice of which is dependent on the overall goals of the experiment and the level of detail desired. Biological networks can then be categorized based upon multiple factors that influence the interaction including directionality, kinetics, strength of the connection between two nodes, and sign (activation or inhibition) of the interaction. Commonly used PPI platforms are summarized in Table 7.1.

Manually curated for both directed (OmniPath and Signor) and undirected (HPRD) resources are typically less detailed and more skewed toward well-characterized protein interactions. In contrast, databases such as BioGRID and BIOPLEX that are constructed from high-throughput experiments, primarily yeast two-hybrid modeling systems, usually contain many more interactions and are more informative. However, these interactions are not individually verified by further experimentation and thus have lower degrees of confidence associated with them. Additional resources such as STRING and HAPPI compile large amounts of interaction data from multiple resources and can be filtered through. However, complications can occur when merging datasets due to an increase in noise, an increase in coverage, and inconsistencies in annotations across datasets.

The three primary steps in constructing PPI networks are as follows:

TABLE 7.1

Overview of Public Resources for Construction of PPI Networks

Resource Name	Description	# of Human Interactions	Species	Weblink
Biomodels	Small-scale dataset for rate-related interactions	10–1,000	Multiple	https://www.ebi.ac.uk/biomodels-main
NFR2ome	Small-scale manually curated oxidative stress database with directed and signed interactions	289	Human	http://nrf2.elte.hu
OmniPath	Manually curated and integrative database for directed and signed signaling	50,247	Human	http://omnipathdb.or
SignaLink2	Manually curated signed and directed signaling interactions and predictions	1,640	Human, *C. elegans,* and *Drosophila*	signalink.org
Signor	Manually curated signed and directed signaling interactions	19,312	Human, mice, rats	signor.uniroma2.it
Reactome	Manually curated pathway database with directed and signed interactions and protein complexes	11,426	Human	www.reactome.org
HPRD	Manually curated undirected interactions	41,327	Human	www.hprd.org
BioPlex	Large-scale mass spectroscopy protein interaction networks	70,000	Human	http://bioplex.hms.harvard.edu
BioGRID	Integrative database of gene and protein interactions from low- and high-throughput publications	406,487	Many species including humans, mice, and rats	https://thebiogrid.org
IntAct	Large-scale collection of protein interactions	310,183	Primarily humans	https://www.ebi.ac.uk/intact
HAPPI	Database of protein interactions with accompanying confidence scores	2,922,202	Human	http://discovery.informatics.uab.edu/HAPPI
STRING	Large-scale predicted and curated interactions	11,353,056	Multiple organisms	https://string-db.org
Mentha	Collection of scored interactions from multiple sources	309,088	Most model organisms	http://mentha.uniroma2.it
InWeb_IM	Collection of PPI datasets with orthological prediction	625,640	Human	http://www.intomics.com/inbio/map/#home

1. Selection of appropriate biological interaction resources.
2. Matching differentially expressed genes to specific proteins.
3. Identify which functions in the PPI can interact with one another.

Tools such as ENRICHNET, NETPEA, and NetWalk use a random walk with restart algorithms approach to perform step 3. In this method, the DEGs are first individually assigned to a protein and the algorithm walks along the interaction network until randomly restarting in order to determine the distance between protein nodes and elucidate their function. The underlying assumption that governs this algorithm is that protein nodes with similar functions will lie within close proximity of each other in an interaction network. This algorithm has previously been used to identify the cell cycle arresting function of p53 in response to sublethal doses of the chemotherapeutic doxorubicin as well as the p53-induced apoptotic response to higher doses of doxorubicin.

An adaptation of the random walk with a restart algorithm is the Ant Colony Optimization method. In this technique, the random walker, which is termed the 'ant', wanders around the interaction network and leaves behind what is referred to as a 'pheromone trail'. After a restart, the pheromone trail increases the probability that the next random walker will traverse the same path as the previous ant. The strength of this pheromone trail is largely dependent on a function of the previously visited protein nodes in the interaction network and connects distant paths into a complex interconnected network.

The following example will use the list of DEGs generated by the DESeq2 Galaxy job for atorvastatin-treated cells discussed in Chapter 3 (p-value cutoff = 0.1). The free-to-use *Reactome* platform was utilized to generate Figure 7.3 and subsequent figures in this section. The following steps summarize the procedure:

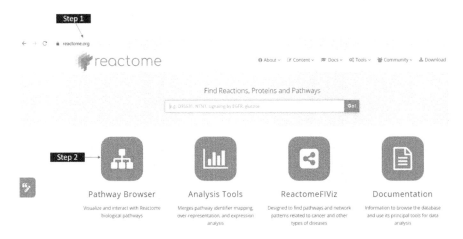

FIGURE 7.3 Steps 1 and 2 of the summarized procedure.

Toxicogenomics and RNA-seq

1. Visit reactome.org
2. Select *Pathway Browser*.
3. In the top-right corner of the interactive resource, select *Analysis*.
4. Under the default *Analyze Gene List* tab, copy and paste the list of DEGs.
5. In the *options* tab, select the default setting of project to human.
6. Click analysis in the lower right corner of the window.
7. Click on statistically significant pathways in order to visualize the mechanisms involved.

Once we have visited www.reactome.org, we are brought to the initial window for the platform and we can then click Pathway browser (Figure 7.3). In addition, we can simply use the search bar to search for pathways by typing in the name of a protein or the name of a signaling pathway directly. Clicking on pathway browser brings us to the main reactome window, giving us access to a variety of different tools. We can change the model organism in which to search for pathways by clicking the scroll down menu near the top of the page as well as observing all of the available pathways in the *event hierarchy* section on the left side of the page. The lower portion of the window is populated by several informatory tabs including gene description (*description*), molecular structures (*structures*), physiologically standard expression levels in different tissue types (*expression*), and the list of pathways (including *p*-values and FDRs) in which our gene list input map to (*analysis*). The false discovery rate (FDR) is an adjusted rate for multiple comparisons that refers to the percentage of null hypotheses that actually hold true.

The next step is to click the analysis option located in the top-right corner of the window (Figure 7.4), which will then take us to the *analysis tools* window. In the analysis gene list tool, we can either upload a file containing our gene

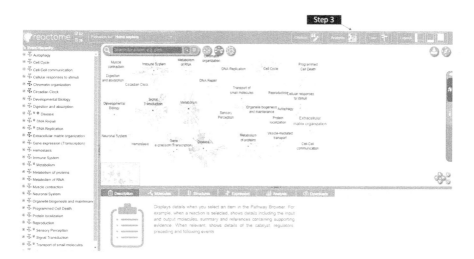

FIGURE 7.4 Step 3 of the summarized procedure.

list of interest or we may choose to directly copy and paste our gene list into the space provided. Other tools available can be accessed on the left side of the menu including analyzing gene expression, species comparison in order to compare orthological pathways across species, and tissue distribution of each gene in the list input. If no data input is available, we may also look at sample gene lists by clicking one of the options underneath the some examples section on the right side of the page.

For our purposes, we will use the analysis gene list tool and copy and paste our list of statistically significant genes generated from DESeq2 for atorvastatin-treated cells (Figure 7.5). After clicking *continue*, we will select project to human under the *options* tab and then click *analyse* once more (Figure 7.6). Based on statistically significant FDR (FDR<0.05), two pathways were identified to be dysregulated: *fructose catabolism* and fructose *metabolism*. Clicking on the pathway name will generate a figure detailing the dysregulated pathway, over which the arrows can be hovered in order to discover information on each individual step. The pathway can be subsequently downloaded as a .pdf file and exported. Figure 7.7 presents the figure generated for the *fructose catabolism* pathway.

7.3.3 CO-EXPRESSION NETWORKS

Co-expression networks are another example of undirected graphs in which each node is representative of a single gene and edges connect genes that share some commonality. In the case of co-expression networks, edges connect genes that demonstrate a greater than expected tendency to have a coordinated expression pattern across multiple samples.

Methods to construct co-expression networks may be either data-driven or knowledge-based and the underlying principle of both methods are similar.

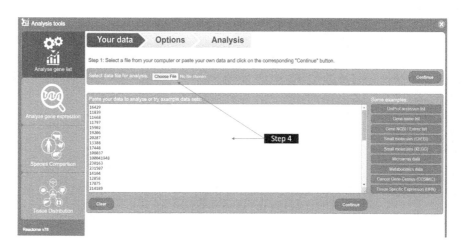

FIGURE 7.5 Step 4 of the summarized procedure.

Toxicogenomics and RNA-seq

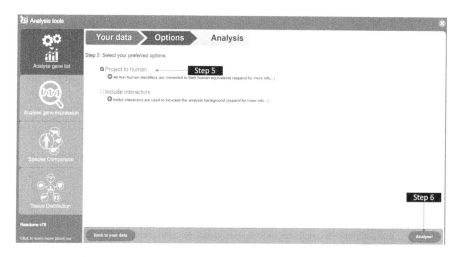

FIGURE 7.6 Steps 5 and 6 of the summarized procedure.

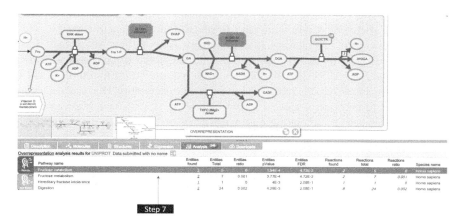

FIGURE 7.7 Step 7 of the summarized procedure. The generated pathway figure for *fructose catabolism* is presented.

There are four primary mathematical measures that are used to construct co-expression networks. These include the following:

1. **Euclidean distance**: A measure of the geometric distance between two vectors (genes/nodes) that takes into account the directionality and magnitude of the vectors (gene expression).
2. **Mutual information (MI)**: A measure of the reduction in uncertainty concerning the expression levels of one gene when the expression levels of another gene are known.

3. **Pearson's correlation coefficient**: A measure of the tendency of the magnitudes of two vectors to exhibit the same sign or direction. This measures how significant the association (co-expression) is between two continuous dependent variables (the two genes).
4. **Spearman's rank correlation**: A modified, nonparametric approach to the Pearson's correlation coefficient that takes into account the rank of gene expression values in a vector. In essence, it assesses how well the relationship between two vectors can be explained by a monotonic function.

Each of the aforementioned measures of association has inherent advantages and disadvantages associated with their use. Euclidean distances should not be utilized if two genes have consistently low expression profiles but are otherwise randomly correlated as they may appear to be close in terms of Euclidean space and yet are distantly related. In addition, Euclidean distances are not appropriate when data contains lots of extraneous variables or is of high dimensionality as the measure is quite sensitive to background noise.

MI is advantageous as it can detect non-linear relationships between nodes; however, these non-linear associations may be too complex and not have biological relevance. Furthermore, quality estimates of the distribution for MI can only be obtained when a relatively large number of samples are used ($n \geq 15$).

Spearman's rank correlation is less sensitive to outliers than Euclidean distance, but also will detect many false positives if the sample size is small. Pearson's correlation coefficient is the most commonly used measure of association for the purposes of generating co-expression networks. Pearson's correlation coefficient takes on a value ranging from -1 to 1 with values closer to -1 indicating a negative correlation and a value closer to one denoting a positive correlation between two genes. In the context of gene expression, a positive correlation value is indicative of an activation mechanism in which the increase in expression of one gene occurs simultaneously with the increase in expression of another gene. In contrast, a negative correlation value occurs when there is an inverse relationship between two genes; that is, the increase in expression of one gene corresponds to a decrease in the expression of the second gene. However, Pearson's correlation coefficient has two primary disadvantages. First, this correlation coefficient can only be used to estimate linear relationships between two genes and this estimate is somewhat sensitive to outliers. The second limitation is that the underlying distribution of gene expression is assumed to follow a normal distribution. Violations of this assumption would thus invalidate the significance of the calculated Pearson correlation.

7.3.3.1 Context Likelihood of Relatedness

There are two primary methods to construct co-expression networks. One of the earliest methods was termed Context Likelihood of Relatedness (CLR), a method of which utilizes estimates of MI between two genes and compares it to a background distribution of MI scores per gene in order to construct similarity networks. One instance of practical use of this network inference algorithm was to

Toxicogenomics and RNA-seq

identify genes transcriptionally regulated by the gene nuclear factor erythroid 2-related factor (Nrf2) under conditions of oxidative stress in mouse lung cells. This study compared the gene findings determined by the CLR method to another MI-based method known as the Algorithm for the Reconstruction of Accurate Cellular Networks (ARACHNE). In both algorithms, the pairwise MI scores are calculated based on the correlations in their gene expression patterns by using the following formula (Eq. 7.1):

$$CLR = \sqrt{Az^2 + Bz^2} \qquad (7.1)$$

In the formula above, Az is the corresponding z-score for the pairwise MI score between the two genes A and B under the MI score distribution of gene A while Bz is the pairwise MI score between the two genes A and B under the MI score distribution of gene B. A greater MI score between two genes is suggestive of greater predictive power to predict the expression pattern (state) of one gene based on the states of the second gene. Thus, a higher MI score between two genes corresponds to a greater likelihood that one gene regulates another. Figure 7.8 provides a sample co-expression network output generated from the CLR algorithm in the aforementioned study.

7.3.3.2 Weighted Gene Co-expression Network Analysis

A second popular algorithm to generate co-expression and PPI networks in response to drug exposure is the Weighted Gene Co-expression Network Analysis (WGCNA). In an unweighted network such as the one used in DGE analysis, a binary scale is used in which 1=connection and 0=no connection. However, it may not be biologically relevant to encode co-expression of genes in such a simplistic manner. Thus, WGCNA assigns a correlation coefficient value that ranges from −1 to 1 in order to describe the strength of the association between two nodes (i.e., how strong the connection is between two genes or gene modules).

WGCNA applies what is termed a 'soft threshold' in order to eliminate noise in the adjacency matrix by raising the correlations to a particular power. The power in which to raise each correlation is calculated based on its similarity to the scale-free graph, a network whose probability distribution follows a power law. A power law is any functional relationship between two nodes where a relative change in the quantity of one node results in a proportional change in the quantity of the second node irrespective of the starting quantities. Although the software that utilizes these algorithms to generate co-expression network is beyond the scope of this textbook, the general steps that are followed include

1. Generation of the co-expression network (correlation matrix),
2. Defining co-expressed gene modules,
3. Relating these modules to external data such as clinical data, Gene Ontology (GO) terms, or pathways.
4. Determination of changed elements between different networks such as a network generated from a control.

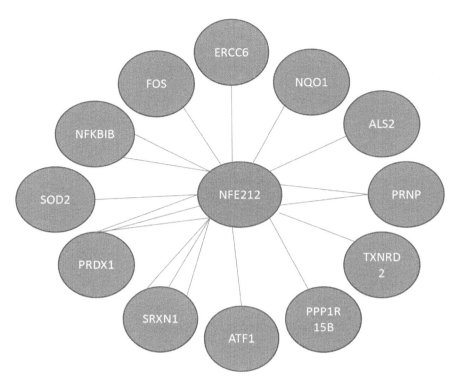

FIGURE 7.8 The genes predicted to be transcriptionally regulated by the transcription factor NFE212 using the CLR algorithm. The Z-score cutoff for significance was set to $z \geq 2.0$ (two standard deviations above the mean).

Establishment of the network in the first step involves the calculation of the correlation coefficients between each gene or probe and then raising that correlation to a soft power to reduce noise. In the next step, modules are defined by construction of a dissimilarity matrix by assessing the topological overlap matrix between nodes, a figure in which compares how similar the two genes are in terms of what they are connected to. If the two genes are connected to many of the same nodes in the network, they would thus have low dissimilar values between them. Once the dissimilarity matrix is completed, hierarchical or k-means clustering can be used to further define these modules and relate these connections to biologically significant external data.

Many practical examples of the use of WGCNA exist and have been published. One such application was analysis of microarray data from mice exposed to the compound chloroprene, a polymer used in the production of plastic (GEO accession #: GSE40795). The samples were divided into mice exposed to noncarcinogenic levels (control) and carcinogenic levels of chloroprene and WGCNA was used to determine significantly different gene modules between the two conditions. Ultimately, 2,434 genes were found to be differentially expressed, of

Toxicogenomics and RNA-seq

which 7 significant hub genes were identified including *Cftr, Hip1, Tbl1x, Ephx1, Cbr3, Antxr2*, and *Ccnd2*. These seven hub genes were mapped to such pathways as the inflammatory response, regulation of gene transcription, cell adhesion, angiogenesis, cell transformation, and cell cycle regulation and were found to be essential in the pathogenesis of lung tissue exposed to chloroprene.

Another application analyzed differential expression and co-expression data in rats with liver fibrosis in order to determine genes associated with liver fibrosis as well as to identify chemical compounds that elicit a similar gene expression profile response. A PPI network was generated from a list of identified genes including *TIMP1, CTGF, TGFB1, MMP-2, LGMN*, and *PLIN3*. These genes were grouped into such modules as wound healing, extracellular matrix organization, and liver cirrhosis. These gene expression profiles were then compared to the profiles of 640 known drug exposures in the public database DrugMatrix and two drugs, carbon tetrachloride and lipopolysaccharides, were found to be associated with early stages of fibrotic damage that were not evident in histopathological samples.

7.3.4 Signature Matching

Once differentially expressed genes are identified between an untreated sample and samples treated with the compound or drug of interest, further protocols can be followed in order to determine whether or not toxicity will occur as well as possible mechanisms that confer the toxicity response. One example is a method known as compound signature matching which has been used extensively in multiple applications including the prediction of side effects based on differential expression profiles as well as for the purposes of drug repurposing.

The defining principle that underlies this procedure is that compounds that induce similar gene expression signatures will induce comparable downstream biological effects. This allows for the prediction of toxicity mechanisms for novel compounds with unknown toxicity profiles by comparing gene expression signatures with known compounds in a database. However, signature matching is typically used to formulate hypotheses for further *in vivo* studies due to differences in cell line responses, timepoints, and dosage responses.

One of the most commonly used *in vitro* databases in which to compare differential expression profiles is connectivity mapping (cMAP). In the cMAP database, a query signature from the test compound, which usually is represented by a list of the most significantly upregulated and downregulated genes, is compared against a reference compound with a known toxicity profile in the database. A connectivity score is then assigned to each comparison in order to determine the strength of the similarities between the two expression profiles.

There are three possible outcomes that can arise due to this method. The first, referred to as positive connectivity, occurs when the differential expression signatures are similar between the test compound query and the reference compound. In this case, a hypothesis could be generated that states the toxicity mechanisms are similar between the two compounds. The second outcome, which is referred to as

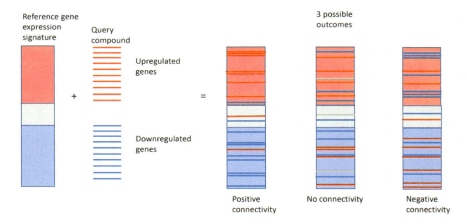

FIGURE 7.9 Illustration of the possible outcomes when comparing the expression signature of a query compound with a reference compound expression signature.

negative connectivity, occurs when there are significant differences in expression signatures between the test compound query and the reference compound such as genes being upregulated in the test query but those same genes being significantly downregulated in the expression signature for the reference compound. In this instance, a preliminary hypothesis would state that there is a significant divergence in toxicity mechanisms between the two compounds. Finally, there can be no connectivity between the two compounds and additional comparisons to other compounds would need to be made. Figure 7.9 shows an example of each possible outcome when performing connectivity mapping for a sample test compound.

7.4 TOXICOGENOMICS DATABASES

Multiple public databases are available that integrate GO terms, pathways, references, and other biological data to aid in predictive toxicology, determination of mode-of-action, and elucidation of adverse outcome pathways. In this section, we will discuss three of the largest and most commonly used databases.

7.4.1 Comparative Toxicogenomics Database

The Comparative Toxicogenomics Database (CTD) was first launched in 2004 and is primarily designed to allow access to manually curated information concerning chemical–gene, chemical–disease, and gene–disease interactions. The database is integrated with Gene Ontology (GO) terms, Kyoto Encyclopedia of Genes and Genomes (KEGG) pathways, and a variety of literature references to allow for hypothesis-driven investigations and exploration of environmentally influenced diseases in both vertebrates and invertebrates. There are eight primary types of data that can be explored and analyzed in the CTD (Figure 7.10) and are presented below:

Toxicogenomics and RNA-seq

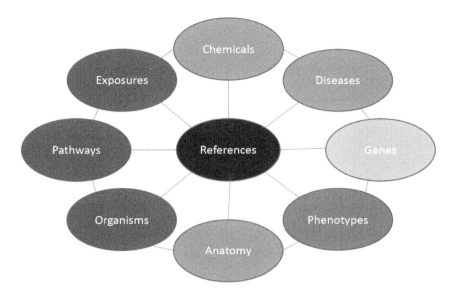

FIGURE 7.10 Overview of the available data in the CTD.

1. **Chemicals**: Integrates the hierarchical vocabulary from the U.S. National Library of Medicine known as the Medical Subject Headings (MeSH). Chemical structures, functional annotations, enriched pathways, curated gene and disease associations, and inferred disease relationships can be explored.
2. **Diseases**: Combines disease descriptors from MeSH and the Online Mendelian Inheritance in Man (OMIM) database. This section is used to curate associations between genes and disease as well as chemicals and disease.
3. **Genes**: Cross-species gene vocabulary including symbols and names is derived from the Gene database of the National Center for Biotechnology Information (NCBI). Curated chemical interactions, curated and inferred disease associations, functional annotations, and important biochemical pathways can be explored and analyzed.
4. **Phenotypes**: In CTD, phenotypes refer to a biological event that is not defined by a particular disease term. For example, abnormal cell cycle arrest is archived as a phenotype while lung cancer is denoted as a disease. Phenotypic vocabulary terms are obtained from the GO. For each chemical, GO terms that are statistically enriched through interaction with a particular chemical are displayed for each chemical.
5. **Anatomy**: Descriptions are derived from the *Anatomy A* branch of MeSH and have a structured format with seven components including chemical, action qualifier, phenotype entity, taxon, anatomy, PubMed reference, and inference network.

6. **Organisms**: Hierarchical organism vocabulary is derived from the Eumetazoa branch of NCBI-based Taxonomy Database.
7. **Pathways**: Known molecular interactions are derived from KEGG and REACTOME pathway data. These pathways are integrated with chemicals, genes, and diseases to allow for insight into possible mechanisms for environmental diseases and allow for hypothesis generation.
8. **Exposures**: Curated exposure data and associations between chemicals and genes.

Chemical–gene/protein interactions, gene–disease associations, chemical–disease associations, chemical–phenotype interactions, and gene–gene interactions are the primary relationships that can be assessed using the CTD. Chemical–gene/protein search inquiries can include both direct associations such as binding of the chemical of interest to a target protein and indirect associations. Interactions in CTD are curated using a hierarchical interaction-type vocabulary system that describes physical, biochemical, and regulatory associations between a chemical and gene/protein.

The search terms are subdivided into three categories. The first category *actions* includes such terms as 'binds to' and describes the interaction event between the chemical and gene/protein of interest. The second category includes *operators* that describe the directionality of the interaction such as 'increases' or 'decreases'. Finally, the third category known as *qualifiers* specifies the form of the gene or chemical that is found in the interaction of interest such as 'chemical metabolite'.

For associations between a gene and a particular disease, CTD contains both curated and inferred associations. Curated associations are taken from either published literature or from the OMIM database while inferred gene–disease associations are noted when the literature has shown separate associations between the chemical and the gene of interest as well as between the chemical and disease.

Figure 7.11 presents an overview of curated and inferred relationships in the CTD. For each inferred relationship between a gene and a disease, an inference

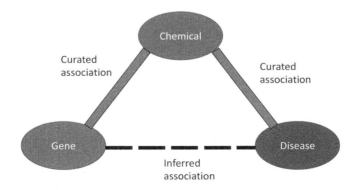

FIGURE 7.11 Overview of the curated and inferred associations in CTD.

Toxicogenomics and RNA-seq   191

score that measures the similarity between a chemical–gene–disease network in CTD to a comparable scale-free random network is calculated with higher scores suggesting significant connectivity (different than expected). Each inference score is a log-adjusted product of two statistics, the first of which measures the connectivity between the chemical and disease while taking into account the number of genes involved and the second of which measures the connectivity between the genes that are used to make the inferential association themselves. Chemical–disease relationships and associated terms operate under a similar range of confines for both curated and inferred associations.

Associations between chemicals and phenotype include eight distinct data entries that are annotated with eight controlled vocabularies. These include the following:

1. **C**: Chemical name taken from CTD chemical vocabulary
2. **Q**: An action qualifier statement that details directionality of the association
3. **E**: Go-derived entity phenotype
4. **A**: MeSH-derived anatomical term
5. **T**: An organism derived from the NCBI taxonomy manual
6. **M**: CTD method code such as *in vivo* or *in vitro*
7. **S**: Information source code
8. **P**: Article identifier for reference (PMID)

Finally, gene–gene interactions in CTD are curated by BioGRID and are derived from primary reference articles.

The following examples will utilize the CTD to assess possible associations between an input gene list and various chemical compounds. The primary window for the CTD (Figure 7.12) can be accessed through the weblink http://ctdbase.org/ and we can use the keyword search bar on the right side of the screen

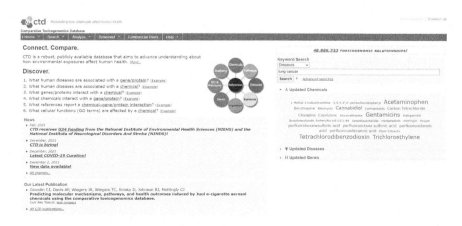

FIGURE 7.12   Main page of the CTD. The search bar can be located on the right side of the window.

to search the database for names of any of the eight previously discussed data categories. For example, we can make a search query for lung cancer by selecting the *diseases* category in the scroll down menu and then inputting our disease of interest (Figure 7.12). Once the search query has been submitted, we can then select our disease of interest from the search output (in this example, non-small cell lung cancer was selected).

The next window is the main information page for our search query (Figure 7.13) and contains cited references as well as the following nine tabs:

1. **Basic**: Contains general information such as name of query, MeSh ID, definition, and hierarchical search ancestry.
2. **Chemical-gene interactions**: Contains name of the chemical, name of the gene, type and directionality of interaction, and references.
3. **Chemicals**: List of chemicals that have either a curated or inferred association with the search query.
4. **Genes**: List of genes that have either a curated or inferred association with the search query.
5. **Phenotypes**: List of phenotypes that have either a gene-GO annotation or inferred association with the search query.
6. **Comps**: In the case of diseases, a list of other diseases that have similar sets of gene and chemical associations when compared to the search query.
7. **Pathways**: List of KEGG and REACTOME pathways that are associated with the search query.
8. **Exposure studies**: List of exposure studies associated with the search query.
9. **Details**: Exposure details as they pertain to the search query.

FIGURE 7.13 Sample information hub for the disease non-small-cell lung carcinoma.

Toxicogenomics and RNA-seq   193

If a particular interaction such as a chemical–gene association is of interest, we can search directly for this association using the *search* scroll down menu and selecting chemical–gene interactions. Figure 7.14 shows a sample search inquiry for the association between the chemical acetaminophen and the gene tumor necrosis factor (TNF) in humans. We can modify the *action* and *operator* terms for our chemical inquiry as well as modify the *qualifier* statement for our gene inquiry. We may also further specify our search inquiry by including the pathway, organism, or GO terms of interest, which may be done either by typing the name of the term or by clicking on the *select* button in order to browse all available terms in CTD. We can finally submit our search inquiry in order to load our search results.

Within the CTD interface, there are three primary tools that we can access for analysis and can be accessed through the *analyze* pull-down menu near the top of the page. The first tool, referred to as the *Set Analyzer* tool, is a four-step process that can be used to perform set-based enrichment analyses for a list of chemicals or genes and is comparable to the previously discussed DAVID platform.

Figure 7.15 displays a sample analysis using the *Set Analyzer* tool. In step 1, we select whether our input list for step 2 will be a list of chemicals or genes. In step 2, we will copy and paste our list of DEGs from the atorvastatin treatment experiment. However, we can instead use any list of genes or chemicals in which we want to identify enriched diseases, GO terms, or pathways. In step 3, we select our preferred analysis and in step 4 we select our cutoff of significance. For our purposes, we will select enriched diseases as the preferred analysis in step 3 and use a corrected *p*-value threshold of 0.05. Figure 7.16 shows a portion of the output list of enriched diseases based on the aforementioned parameters. Such disease pathways as skin diseases, gastrointestinal diseases, and neoplasms of the lungs and gastrointestinal tract were found to be enriched based on the gene list input.

The second useful tool available in the CTD is termed *MyGeneVenn* and allows us to generate a Venn diagram to assess the similarities and differences

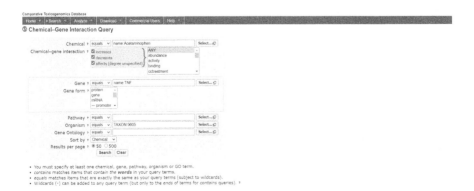

FIGURE 7.14 A sample search inquiry for the association between the chemical acetaminophen and the gene TNF in humans.

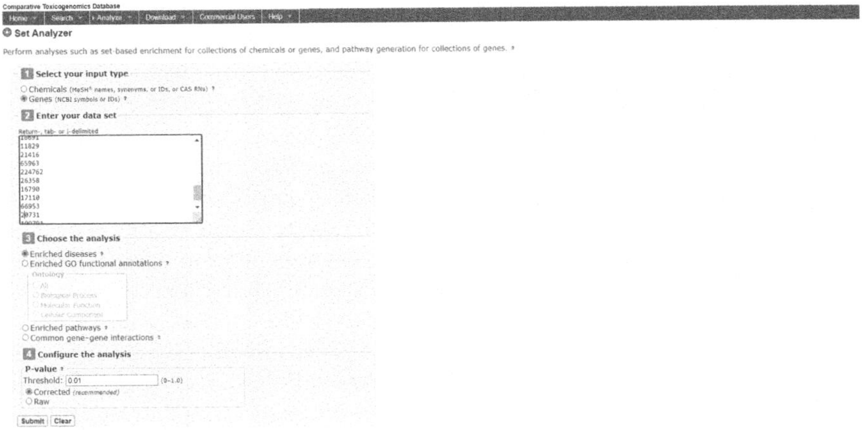

RNA-seq in Drug Discovery and Development

FIGURE 7.15 Sample analysis using the *Set Analyzer* tool in CTD. The DEGs list from the atorvastatin example was used for the dataset.

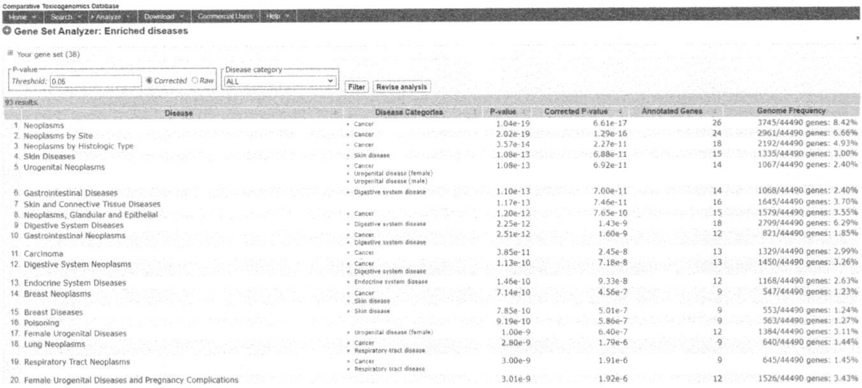

FIGURE 7.16 Top 20 enriched diseases generated from the *Set Analyzer* tool in CTD.

between our inputted gene list and the genes associated with up to two chemicals or diseases. In step 1, we will copy and paste the same list of DEGs as used for the previous example using the *Set Analyzer* tool. In addition, the name of the Venn diagram can be modified in this step. For step 2, the three following comparisons can be made:

1. genes from the list that interact with up to two chemicals,
2. genes from the list that have curated associations with up to two diseases, and
3. genes from the list that have inferred associations with up to two diseases.

Toxicogenomics and RNA-seq 195

Finally, step 3 allows us to decide whether we want to only include direct relationships or to also include any descendent terms.

In order to avoid having too many hits in our output, we will select the first option in which only direct relationships are included. Figure 7.17 presents an example using the first comparison between the list genes and interactions with the two chemicals acetaminophen and ethylene dichloride. From this example, 3 of the 38 genes from our DEG list were not associated with either chemical, 1 gene was associated with ethylene dichloride only, 19 genes were associated with acetaminophen only, and 15 genes were associated with both chemicals. The list of 15 genes that are associated with both ethylene dichloride and acetaminophen is also presented in Figure 7.17. Figure 7.18 presents an example using the same gene list as the one used in Figure 7.17, but instead we will change step 2 to include a list of genes with curated associations to the two diseases acute kidney injury (AKI) and hepatocellular carcinoma. From this example, 33 of the 38 genes from our DEG list were not associated with either disease, 3 genes were associated with hepatocellular carcinoma only, 1 gene was associated with AKI only, and 1 gene was associated with both diseases.

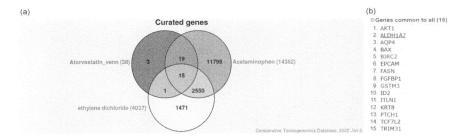

FIGURE 7.17 Venn diagram generated from shared genes between gene list and the chemicals acetaminophen and ethylene dichloride. The 15 genes associated with both chemicals are also presented.

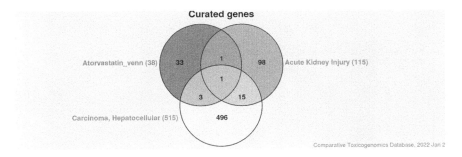

FIGURE 7.18 Venn diagram generated from shared genes between gene list and the diseases hepatocellular carcinoma and AKI.

The third tool that is available is the *MyVenn* tool, which allows us to assess overlaps in information and other associations between different datasets. In step 1, we select the input type of our datasets with available options including chemicals, disease names, genes, GO terms, or pathway names from KEGG/REACTOME. For our example, we will compare the list of differentially expressed genes between cells treated with atorvastatin and cells treated with albendazole in order to identify common genes that are differentially expressed in both conditions. In step 2, we input our separate DEG lists for atorvastatin-treated and albendazole-treated into *set1* and *set2*, respectively, and submit the query (Figure 7.19). As shown in Figure 7.20, 13 genes were found to be differentially expressed in both atorvastatin-treated and albendazole-treated cells.

7.4.2 Japanese Toxicogenomics Project

The Japanese Toxicogenomics Project (TGP) was released for free public use in 2011 and it compiles thousands of microarray studies from a collaborative effort made by the Japanese National Institute of Health Science, the National Institute

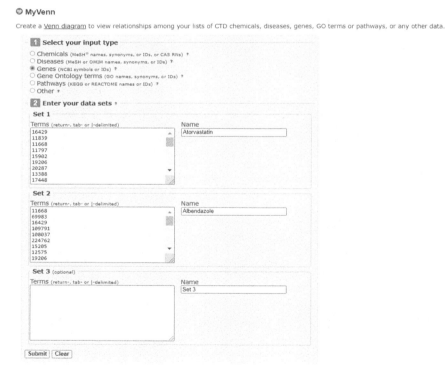

FIGURE 7.19 Parameters to generate a Venn diagram comparing DEG between atorvastatin-treated and albendazole-treated cells using the *MyVenn* tool in CTD.

Toxicogenomics and RNA-seq

FIGURE 7.20 Venn diagram showing shared DEGs between atorvastatin-treated and albendazole-treated cells.

Biomedical Innovation, and 15 different pharmaceutical companies. In particular, the TGP focuses on the gene expression profiles of liver cells in response to 170 different marketed medicinal drugs with some supplementation of expression profiles derived from kidney cells.

The TGP, like the public database DrugMatrix which will be discussed in the next section, is advantageous when compared to less toxicogenomics-focused databases like ArrayExpress and GEO for three primary reasons:

1. The uniform experimental design allows for easier and more comprehensive comparisons between treatments.
2. The wide use of marketed and human-relevant drugs allows for extensive research concerning predictive toxicology models.
3. The simultaneous use of both *in vivo* and *in vitro* studies allows for assessment of the similarities and differences between the types of studies in predictive toxicology.

In vivo experiments were performed on male Sprague Daly rats with two different setups, one of which was a single-dose study and the other of which was a repeated-dose study. For each of the two experimental designs, a separate one-week study was performed in order to identify the maximum tolerated dose (MTD) for the 170 compounds. The MTD was then set as the largest dose in both the single-dose and repeated-dose experiments, each of which was done in triplicates. For the single-dose experiment, rats were treated with either a low, medium, or high dose in a 1:3:10 ratio and were sacrificed at 3, 6, 9, or 24 hours. Rats in the repeated-dose experiment were also administered a low, medium, or high dose in a 1:3:10 ratio, but were subsequently sacrificed 24 hours after the last dose was administered at 3, 7, 14, and 28 days.

Beyond gene expression data, histological examinations, blood chemistry, body weight, organ weight, hematology, and physiological symptoms were also collected. In addition to *in vivo* studies, primary hepatocytes were treated with a low, medium, or high dose (1:5:25 ratio) and were harvested for gene expression analysis at 2, 8, and 24 hours post-treatment. Table 7.2 summarizes the key components of the TGP database.

198   RNA-seq in Drug Discovery and Development

TABLE 7.2

Overview of the Important Background Characteristics of the Experimental Design Used to Collect Samples Populating the TGP Database

Species	Male Sprague Dawley rat	Male Sprague Dawley rat	Male Sprague Dawley rat
Study design	*In vivo*	*In vivo*	*In vitro*
Dose type	Single dose	Repeat dose	Single dose
Dose levels	Control, low, medium, and high	Control, low, medium, and high	Control and high
Sample collection intervals	3, 6, 9, and 24 hours	24 hours post-treatment at 3, 7, 14, and 28 days	16 and 24 hours
# of Replicates	Triplicates	Triplicates	Duplicates
Microarray platform	Affymetrix RG230–2.0	Affymetrix RG230–2.0	Affymetrix RG230–2.0
Toxicology information	Body weight, organ weight, histopathology, blood chemistry, and food consumption	Body weight, organ weight, histopathology, blood chemistry, and food consumption	Cell viability

7.4.3 DrugMatrix

Like TGP, DrugMatrix contains genomic data and typical toxicology data derived from both *in vivo* and *in vitro* studies in male Sprague Dawley rats. For the *in vivo* experiments, two different doses were utilized. The first dose was the fully effective dose, which was defined as the typical treatment dose for disease converted from the human dose, while the second was the MTD, which was defined as a 50% decrease in weight gain relative to the control after 5 days of regular dosing. Most microarray analysis was performed at 6-hour, 1-day, 3-day and 5-day intervals; however, some were switched to 7/14/28/90-day intervals. The GE CodeLink RU1 10,000 rat array system was used for all samples while some samples used the Affymetrix RG230–2.0 arrays as supplementation. Additional toxicology information from serum chemistry tests, hematology, organ weight, and histopathology were also collected. *In vitro* experiments were also performed using primary hepatocytes isolated from male Sprague Dawley rats and gene expression analysis was performed 16 and 24 hours after treatment. *In vivo* studies were performed in triplicates, while *in vitro* studies were performed in duplicates.

DrugMatrix shares many similarities with TGP, allowing for extensive cross-lab comparisons and meta-analyses with regard to comparisons in technical performance, the applicability of genomic markers across studies, and the ability to

Toxicogenomics and RNA-seq

statistically validate findings. The primary commonalities shared between TGP and DrugMatrix are as follows:

1. Focus on gene expression responses due to drug treatment.
2. Data is collected primarily from liver cells (target organ).
3. Significant supplementation with data derived from the kidneys.
4. The same affymetrix Rat RG230–2.0 array chip was used.
5. Conventional toxicology information such as histopathology, blood chemistry, body weight, and organ weight were collected to aid in predictive modeling.
6. Assayed primarily marketed drugs (73 different drugs were assayed in both databases).
7. Used male Sprague Dawley rats.
8. Similar experimental designs were used for both *in vivo* and *in vitro* studies.

However, some key differences exist between the two databases. First, while both databases used the same affymetrix platform for microarray analysis, the affymetrix platform was used for all samples in TGP yet only a portion of the samples contained in DrugMatrix. Microarray analysis in DrugMatrix primarily used the CodeLink platform. Second, the definition used for MTD was distinct between the two databases. The definition for samples derived from DrugMatrix was more precise in that it was determined by what dose was required for a 50% decrease in weight gain. Table 7.3 summarizes the major components of the DrugMatrix database.

TABLE 7.3
Overview of the Important Background Characteristics of the Experimental Design Used to Collect Samples Populating the DrugMatrix Database

Species	Male Sprague Dawley rat	Male Sprague Dawley rat
Study design	In vivo	*In vitro*
Dose type	Repeat dose	Single dose
Dose levels	Control, low, and high	Control and high
Sample collection intervals	0.25, 1, 3, and 5 days	16 and 24 hours
# of Replicates	Triplicates	Duplicates
Microarray platform	Affymetrix RG230–2.0/Codelink RU1 assays	Affymetrix RG230–2.0
Toxicology information	Body weight, organ weight, histopathology, blood chemistry, and food consumption	N/A

7.5 COMPARING MICROARRAY VS. RNA-seq

The final section of this chapter will compare the efficacy of microarray and RNA-seq prediction modeling in toxicogenomics. Toxicogenomics is a discipline concerned with the elucidation of the underlying genomic responses and mechanisms that arise from toxic exposure. Pharmacogenomics, a closely related discipline, is concerned with the identification of inter-individual variations in the whole genome as well as specific nuances in individual candidate genes including single-nucleotide polymorphisms, alterations in gene expression, and haplotype markers that influence the severity of a xenobiotic response. In essence, the field of pharmacogenomics is concerned with identifying alleles and other genomic features that can increase or decrease an individual's response to toxin or drug exposure. This knowledge of patient-specific expression levels or allelic differences in critical genes allows for more individualized treatment plans, reduced risk of adverse drug reactions, and optimization of drug efficacy.

The relatively new fields of toxicogenomics and pharmacogenomics have largely been developed through the use of microarray technology with regard to assessing gene expression. In 2011, the Microarray Quality Control consortium, which was led and funded primarily by the Food and Drug Administration (FDA), was launched and performed a nearly 4-year study in order to compare the efficacy of microarrays and RNA-seq in patient outcome prediction and biomarker transferability across platforms. We will first discuss their findings with respect to patient outcome prediction.

The use of microarrays and RNA-seq is prevalent in the study of cancer. The neuroblastoma patient outcome prediction portion of the FDA-led study used both microarrays and RNA-seq to generate expression data for 498 primary neuroblastoma samples in order to compare the endpoint prediction capabilities of both techniques. RNA-seq data generated a list of 48,415 unique transcript ids that were differentially expressed while traditional microarray data only detected 21,101 unique transcripts. Of the 48,415 unique transcripts identified through RNA sequencing, 34,175 transcripts corresponded to coding genes while the remaining 14,240 genes corresponded to non-coding genes. 360 predictive models were subsequently generated by first dividing the 498 primary neuroblastoma samples into training and validation sets. The following six endpoints were used to build the prediction models:

1. patient sex (SEX),
2. event-free survival among all patients (EFS_all),
3. overall survival among all patients (OS_all),
4. favorable disease outcome (FAV),
5. event-free survival for high-risk patients (EFS_HR), and
6. overall survival for high-risk patients (OS_HR).

Toxicogenomics and RNA-seq

Despite RNA-seq providing more transcriptomic data than the microarray, the performance of the prediction models was comparable across platforms and largely depended on the clinical endpoints utilized in the model.

The next component of the study focused on assessing whether or not RNA-seq was better at predicting and determining toxicity mechanisms in response to a variety of compounds. Twenty-seven test chemical compounds representing seven modes of action for toxicity responses with triplicates were performed for each chemical. As was the case with the neuroblastoma experiment presented above, samples were divided into training and test sets and models were generated from both RNA-seq and microarray data.

Results from the study showed that treatment effect greatly influenced the concordance of the data between the two platforms with compounds associated with low treatment effect showing less cross-platform consistency than compounds with large treatment effects. In addition, DEGs that were lowly expressed showed greater inconsistencies between microarray and RNA-seq methodology due to the decrease in sensitivity in microarrays. The overall predictive accuracy for models generated from microarray and RNA-seq data was 58% and 61%, respectively. Thus, RNA-seq may be a more efficient and comprehensive platform to use, especially when the treatment effect for a compound of interest is relatively minor or the target genes of interest are lowly expressed.

The last investigation arm of the SEQC consortium, referred to as the gene signature transferability investigation, focused on the viability and applicability of predictive models and biomarkers generated from microarray data to RNA-seq data. Three human and two rat datasets were utilized with three mapping complexities identified that were used to group genes. In order of increasing complexity, the three modeling complexities that were used were k-nearest neighbors, nearest shrunken centroids, and support vector machine.

Using these three parameters, approximately 240,000 different models were ultimately assessed. Signature genes derived from one platform (RNA-seq or microarray) could be used to generate models from the other platform with minimal loss of predictive power. However, the application of predictive models directly across platforms demonstrated mixed results with microarray-based predictive models being able to more accurately predict RNA-seq samples than RNA-seq predictive models with regard to predicting microarray-derived samples. Figure 7.21 provides an overview of the sources of transcriptomic data, experimental methodologies, and inferential toxicogenomics. Thus, despite the ongoing advancements of RNA-seq technology, microarrays still play a significant role in toxicogenomics.

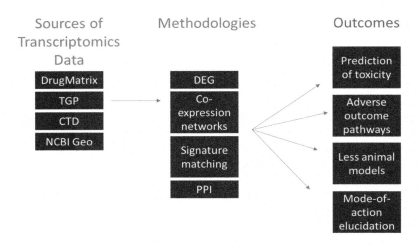

FIGURE 7.21 Overview of the sources of transcriptomic data, experimental methodologies, and inferential toxicogenomics.

7.6 SUMMARY

In this chapter, we introduced the field of toxicogenomics and discussed how this area of study differs from phenotype-based toxicology studies which relied on observable endpoints. Next, we discussed the major types of toxicogenomic analysis including differentially expressed genes, co-expression networks, and signature matching. An example using the reactome platform and our list of DEGs from the atorvastatin treatment example was performed in a stepwise manner in order to generate a PPI. This was followed by the discussion of publicly available toxicogenomic databases such as DrugMatrix and TGP. Finally, this chapter compared the efficacy of RNA-seq and microarray-based data in toxicogenomics and assessed whether microarray data still played a role in the toxicogenomics field.

KEYWORDS AND PHRASES

After reading this chapter, you should be able to demonstrate familiarity with the following words and phrases:

- Understand how traditional toxicity studies were conducted.
- Understand toxicogenomics and its advantages and weaknesses.
- Understand and be able to describe the major toxicogenomic studies including differentially expressed genes, gene networks, co-expression networks, and signature matching.
- Be able to identify and access public toxicogenomic databases.

BIBLIOGRAPHY

AbdulHameed, M. D., Tawa, G. J., Kumar, K., Ippolito, D. L., Lewis, J. A., Stallings, J. D., & Wallqvist, A. (2014). Systems level analysis and identification of pathways and networks associated with liver fibrosis. *PLoS One, 9*(11), e112193. https://doi.org/10.1371/journal.pone.0112193

Alexander-Dann, B., Pruteanu, L. L., Oerton, E., Sharma, N., Berindan-Neagoe, I., Modos, D., & Bender, A. (2018, Aug 6). Developments in toxicogenomics: Understanding and predicting compound-induced toxicity from gene expression data. *Mol Omics, 14*(4), 218–236. https://doi.org/10.1039/c8mo00042e

Boverhof, D. R., & Zacharewski, T. R. (2006, Feb). Toxicogenomics in risk assessment: Applications and needs. *Toxicol Sci, 89*(2), 352–360. https://doi.org/10.1093/toxsci/kfj018

Chen, M., Zhang, M., Borlak, J., & Tong, W. (2012, Dec). A decade of toxicogenomic research and its contribution to toxicological science. *Toxicol Sci, 130*(2), 217–228. https://doi.org/10.1093/toxsci/kfs223

Fabregat, A., Sidiropoulos, K., Viteri, G., Forner, O., Marin-Garcia, P., Arnau, V., D'Eustachio, P., Stein, L., & Hermjakob, H. (2017, Mar 2). Reactome pathway analysis: A high-performance in-memory approach. *BMC Bioinformatics, 18*(1), 142. https://doi.org/10.1186/s12859-017-1559-2

Federico, A., Serra, A., Ha, M. K., Kohonen, P., Choi, J. S., Liampa, I., Nymark, P., Sanabria, N., Cattelani, L., Fratello, M., Kinaret, P. A. S., Jagiello, K., Puzyn, T., Melagraki, G., Gulumian, M., Afantitis, A., Sarimveis, H., Yoon, T. H., Grafstrom, R., & Greco, D. (2020, May 8). Transcriptomics in toxicogenomics: Preprocessing and differential expression analysis for high quality data. *Nanomaterials (Basel), 10*(5), 11–15. https://doi.org/10.3390/nano10050903

Griss, J., Viteri, G., Sidiropoulos, K., Nguyen, V., Fabregat, A., & Hermjakob, H. (2020, Dec). ReactomeGSA - Efficient Multi-Omics Comparative Pathway Analysis. *Mol Cell Proteomics, 19*(12), 2115–2125. https://doi.org/10.1074/mcp.TIR120.002155

Guo, Y., & Xing, Y. (2016, Apr 15). Weighted gene co-expression network analysis of pneumocytes under exposure to a carcinogenic dose of chloroprene. *Life Sci, 151*, 339–347. https://doi.org/10.1016/j.lfs.2016.02.074

He, F., & Maslov, S. (2016, Dec 16). Pan- and core- network analysis of co-expression genes in a model plant. *Sci Rep, 6*, 38956. https://doi.org/10.1038/srep38956

Igarashi, Y., Nakatsu, N., Yamashita, T., Ono, A., Ohno, Y., Urushidani, T., & Yamada, H. (2015, Jan). Open TG-GATEs: A large-scale toxicogenomics database. *Nucleic Acids Res, 43*(Database issue), D921–D927. https://doi.org/10.1093/nar/gku955

Jassal, B., Matthews, L., Viteri, G., Gong, C., Lorente, P., Fabregat, A., Sidiropoulos, K., Cook, J., Gillespie, M., Haw, R., Loney, F., May, B., Milacic, M., Rothfels, K., Sevilla, C., Shamovsky, V., Shorser, S., Varusai, T., Weiser, J., Wu, G., Stein, L., Hermjakob, H., & D'Eustachio, P. (2020, Jan 8). The reactome pathway knowledgebase. *Nucleic Acids Res, 48*(D1), D498–D503. https://doi.org/10.1093/nar/gkz1031

Kinaret, P. A. S., Serra, A., Federico, A., Kohonen, P., Nymark, P., Liampa, I., Ha, M. K., Choi, J. S., Jagiello, K., Sanabria, N., Melagraki, G., Cattelani, L., Fratello, M., Sarimveis, H., Afantitis, A., Yoon, T. H., Gulumian, M., Grafstrom, R., Puzyn, T., & Greco, D. (2020, Apr 15). Transcriptomics in toxicogenomics: Experimental design, technologies, publicly available data, and regulatory aspects. *Nanomaterials (Basel), 10*(4), 51–68. https://doi.org/10.3390/nano10040750

Krewski, D., Acosta, D., Jr., Andersen, M., Anderson, H., Bailar, J. C., 3rd, Boekelheide, K., Brent, R., Charnley, G., Cheung, V. G., Green, S., Jr., Kelsey, K. T., Kerkvliet, N. I., Li, A. A., McCray, L., Meyer, O., Patterson, R. D., Pennie, W., Scala, R. A.,

Solomon, G. M., Stephens, M., Yager, J., & Zeise, L. (2010, Feb). Toxicity testing in the 21st century: A vision and a strategy. *J Toxicol Environ Health B Crit Rev, 13*(2–4), 51–138. https://doi.org/10.1080/10937404.2010.483176

Mattingly, C. J., Rosenstein, M. C., Colby, G. T., Forrest, J. N., Jr., & Boyer, J. L. (2006, Sep 1). The comparative toxicogenomics database (CTD): A resource for comparative toxicological studies. *J Exp Zool A Comp Exp Biol, 305*(9), 689–692. https://doi.org/10.1002/jez.a.307

Reverter, A., & Chan, E. K. (2008, Nov 1). Combining partial correlation and an information theory approach to the reversed engineering of gene co-expression networks. *Bioinformatics, 24*(21), 2491–2497. https://doi.org/10.1093/bioinformatics/btn482

Ritchie, M. E., Phipson, B., Wu, D., Hu, Y., Law, C. W., Shi, W., & Smyth, G. K. (2015, Apr 20). *Limma* powers differential expression analyses for RNA-sequencing and microarray studies. *Nucleic Acids Res, 43*(7), e47. https://doi.org/10.1093/nar/gkv007

Stuart, J. M., Segal, E., Koller, D., & Kim, S. K. (2003, Oct 10). A gene-coexpression network for global discovery of conserved genetic modules. *Science, 302*(5643), 249–255. https://doi.org/10.1126/science.1087447

Taylor, R. C., Acquaah-Mensah, G., Singhal, M., Malhotra, D., & Biswal, S. (2008, Aug 29). Network inference algorithms elucidate Nrf2 regulation of mouse lung oxidative stress. *PLoS Comput Biol, 4*(8), e1000166. https://doi.org/10.1371/journal.pcbi.1000166

van Dam, S., Vosa, U., van der Graaf, A., Franke, L., & de Magalhaes, J. P. (2018, Jul 20). Gene co-expression analysis for functional classification and gene-disease predictions. *Brief Bioinform, 19*(4), 575–592. https://doi.org/10.1093/bib/bbw139

Van Norman, G. A. (2019, Nov). Limitations of animal studies for predicting toxicity in clinical trials: Is it time to rethink our current approach? *JACC Basic Transl Sci, 4*(7), 845–854. https://doi.org/10.1016/j.jacbts.2019.10.008

Waters, M. D., & Fostel, J. M. (2004, Dec). Toxicogenomics and systems toxicology: Aims and prospects. *Nat Rev Genet, 5*(12), 936–948. https://doi.org/10.1038/nrg1493

Xiao, J., Blatti, C., & Sinha, S. (2018, Jul 1). SigMat: A classification scheme for gene signature matching. *Bioinformatics, 34*(13), i547–i554. https://doi.org/10.1093/bioinformatics/bty251

Xu, J., Gong, B., Wu, L., Thakkar, S., Hong, H., & Tong, W. (2016, Mar 15). Comprehensive assessments of RNA-seq by the SEQC consortium: FDA-led efforts advance precision medicine. *Pharmaceutics, 8*(1), 47–55. https://doi.org/10.3390/pharmaceutics8010008

Zhang, B., & Horvath, S. (2005). A general framework for weighted gene co-expression network analysis. *Stat Appl Genet Mol Biol, 4*, 17. https://doi.org/10.2202/1544-6115.1128

8 Herbal Medicine and RNA-seq

Robert Morris and Feng Cheng
University of South Florida

CONTENTS

8.1 What Is Herbal Medicine? 205
    8.1.1 Traditional Medicine.. 205
    8.1.2 Herbal Medicine 206
    8.1.3 Use of Database for Bioactive Compound Example 208
    8.1.4 Properties of Candidate Herbal Compounds 209
    8.1.5 RNA-seq and Herbal Medicine 210
8.2 Mining Functional Genes of Medicinal Plants 211
8.3 Discovery of Secondary Metabolites and Their Metabolic Pathways 214
8.4 Discovery of Developmental Mechanisms 218
8.5 Development of Molecular Markers to Improve Plant Breeding............. 219
8.6 Identification of Target Genes and Molecular Mechanisms of Herbal
    Drugs... 220
8.7 Synergism of Herbal Compounds in Pathway Regulation..................... 222
8.8 Herbal Medicine Toxicity 223
8.9 Natural Drug Repurposing... 226
8.10 Summary 226
Keywords and Phrases 227
Bibliography 227

8.1 WHAT IS HERBAL MEDICINE?

8.1.1 TRADITIONAL MEDICINE

Traditional medicine, which also encompasses traditional Chinese medicine (TCM), refers to the amalgam of medical practices and beliefs that incorporate animal-based and plant-based medicine as well as spiritual and physical exercises. These techniques and remedies may be applied exclusively but are usually paired with conventional medicine in order to prevent illness, treat ailments, and maintain a state of good health. In many cultures, particularly poorer nations with restricted access to conventional medicine and therapeutics, as much as 80% of the general population relies primarily on alternative medicine practices for healthcare. In addition, various forms of alternative medicine are becoming

DOI: 10.1201/9781003174028-8

increasingly popular due to the rise of antibiotic-resistant strains and the steady increase in drug costs. In many instances, extracts are derived from natural products (NPs) and used for a variety of purposes such as to decrease inflammation, improve circulation, or alleviate pain.

8.1.2 Herbal Medicine

Herbal medicine is a critical component of traditional medicine that focuses on the use of herbal and plant components to promote a state of overall health physically, mentally, and spiritually. A holistic approach is used to assess and diagnose a patient and is governed by a set of principles that differ depending on the geographic and cultural background of the practice. For instance, TCM seeks to establish a balanced equilibrium between the concepts of yin and yang in order to alleviate symptoms and ailments. Due to the rapid increase in the average life expectancy, the global use of TCM and other herbal remedies has risen to combat the growing prevalence of chronic complications including arthritis, immune deficiencies associated with aging, and deterioration of bone and muscle function. In addition, herbal medicine usage has been shown to increase where current allopathic medicine has failed to treat a given symptom or disease such as certain types of high-stage cancers.

Approximately 10,000 different medicinal herb species are utilized in TCM, each of which has complex molecular profiles. Medicinal herb and plant species can be processed and administered in many forms including powders, teas, essential oils, capsules, and ointments. Plants commonly used for herbal medicine are comprised of many aromatic and natural components. These plant-derived compounds may be used directly for therapeutic use or may serve as a basis for the development of pharmaceuticals. Termed ethnobotanicals, the diversity of these compounds may be more effective in treating certain disease states than the one-drug/one-target/one-therapeutic approach frequently utilized by allopathic medicine. Conversely, the complexity of many herbal remedies and the wide array of natural compounds may also increase the complexity of adverse reactions and side effects. In addition, herbal remedies are frequently taken alongside various pharmaceuticals, creating the possibility of cross reactions between the supplement and the prescribed drug. Thus, the expansion of a systems-biology approach to analyzing the efficacy of herbal treatments and the mechanisms in which they confer their beneficial effects is necessary.

Chemical compound information is primarily derived from three sources:

1. Separation and purification of compounds (classical pharmacology);
2. Review of the literature; and
3. Small molecule compound databases.

Many different software and public databases are available for TCM research and some of the most popular databases are summarized in Table 8.1. Effective medicinal TCM databases include the following information for each compound:

TABLE 8.1
Public Databases for Herbal Medicine

Database	Organization	Type	Overview
PubChem	US National Institute of Health (NIH)	Compound and drug information	Includes physical and chemical properties as well as biological test results for over 700,000 unique compounds
TCM Database@ Taiwan	China Medical University, Taiwan	TCM	Includes physical and chemical properties for approximately 500 TCMs and over 20,000 ingredients as well as 3D structures.
Chem-TCM	Institute of Pharmaceutical Science at King's College, UK	TCM	Includes chemical and physical properties for approximately 350 TCMs and 9,500 compound as well as target information
AGRICOLA	US Department of Agriculture	Agricultural and bibliographical	Citation database containing over 6 million references to journal manuscripts, theses, presentation recordings, and patents pertaining to botany, agriculture, and related disciplines
HerbMed	Alternative Medicine Foundation	Alternative medicine and bibliographical	Electronic citation database with references to scientific studies related to the efficacy and use of herbal components
Dr Duke's Phytochemical and Ethnobotanical Databases	US Department of Agriculture	Plant and herbal compounds	Large database containing detailed information on bioactivity, toxicity, and chemical function of many herbs, plants, and compounds

1. Compound storage number
2. Name of the compound and its CAS ID
3. Sources of plant information including Latin name and extractive fraction
4. Compound structure such as SMILES code or InChiKey
5. Relative molecular weight
6. The number of rotatable bonds
7. Additional physical and chemical properties

While many databases exist that provide information concerning bioactive compounds and potential interactions, many databases are specialized for one particular aspect of herbal drug discovery. Databases like the TCM Database@ Taiwan and Chem-TCM categorize plant-derived compounds based on the principles of TCM. However, Chem-TCM is not free to use for publishing purposes and TCM Database@Taiwan is somewhat outdated and has some redundant data entries. PubChem is a large dictionary containing physical and chemical properties for a variety of drugs and bioactive compounds. AGRICOLA and HerbMed are useful portals to use initially to find resources and evidence-based support for various treatment paradigms. Finally, Dr Duke's Phytochemical and Ethnobotanical Databases supported by the US Department of Agriculture provide important information regarding structure, activity, and possible side effects for many herbal treatments, plant compounds, and available supplements. Because no single database contains all information concerning a given compound, it is often best to combine the results and information obtained from multiple databases.

8.1.3 Use of Database for Bioactive Compound Example

In the following example, we will utilize the free-to-use PubChem database to find information concerning the compound ginsenol, a tertiary alcohol that is one of the key bioactive compounds in the herbal supplement ginseng. PubChem can be used to search for chemicals and bioactive compounds by their name, molecular formula, structure, drug class, or target genes and can be accessed using the link https://pubchem.ncbi.nlm.nih.gov/. This link subsequently takes us to the main search engine page for the PubChem database (Figure 8.1), at which point we can search for our compound of interest (ginsenol). Once we submit the search inquiry, the database will list the compounds that best match with our search parameters. We then are provided with a large catalog of information associated with our search compound including 2D-structure (Figure 8.2a), 3D-structure

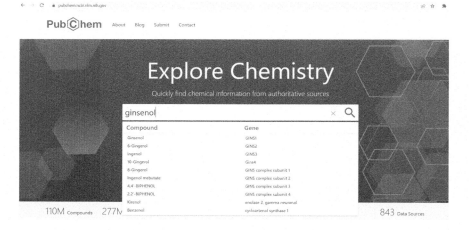

FIGURE 8.1 Main window for the PubChem database.

Herbal Medicine and RNA-seq

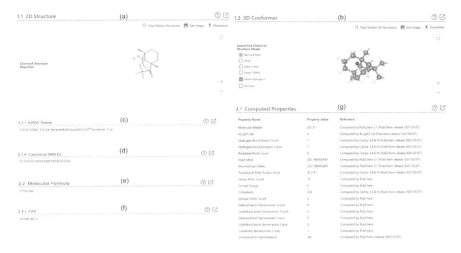

FIGURE 8.2 Sample information for the compound ginsenol using the PubChem database including 2D-structure (a), 3D-structure (b), the International Union of Pure and Applied Chemistry nomenclature (c), canonical SMILES that represent a 1D string of the 2D-structure (d), molecular formula (e), CAS identifier (f), and physical and chemical properties such as molecular weight (g).

(Figure 8.2b), the International Union of Pure and Applied Chemistry (IUPAC) nomenclature (Figure 8.2c), canonical SMILES that represent a 1D string of the 2D-structure (Figure 8.2d), molecular formula (Figure 8.2e), CAS identifier (Figure 8.2f), and physical and chemical properties such as molecular weight (Figure 8.2g). In addition, you can access such information as structurally related compounds, mass spectrometry figures, and literature sources for gene–chemical co-occurrences. Finally, we also are able to look at bioassays that assess bioactivity of the compound of interest. For example, one study found an antifungal activity against the fungal species *Botrytis cinerea* at a dose of 100 ppm for 3 days.

8.1.4 Properties of Candidate Herbal Compounds

Large numbers of compounds can be purified from medicinal plant specimens. However, many of these compounds have little therapeutic relevance due to low pharmacological potency. In order to improve screening efficiency and eliminate compounds that are highly unlikely to yield effective therapeutics, we first test the drug-likeness of each compound. This allows us to predict which compounds have the highest potential and speed up the process of drug-testing as opposed to hand-testing each compound individually for efficacy. Assessment of drug-like qualities is a qualitative approach that looks to assess the likelihood of a natural compound in being an effective candidate for drug synthesis based on structural and chemical properties. In general, the following properties are of concern when

210 RNA-seq in Drug Discovery and Development

determining possible drug candidates, particularly orally administered drugs, that should be analyzed further:

1. Structural characteristics including the number of hydrogen bonds, polarity strength, molecular weight, and shape.
2. Physicochemical properties including pH, solubility, and chemical stability.
3. Biochemical properties including binding affinity and metabolic characteristics.
4. Pharmacokinetic and toxicity properties including lethal dose (LD50), bioavailability, and known drug interactions.

A general rule known as Lipinski's rule of five is composed of a series of criteria that are frequently used to assess whether or not a small molecule is suitable for subsequent drug development. Note that these criteria are not all inclusive and there are many drug exceptions that may not strictly adhere to these parameters. However, it is a useful guideline in filtering out small molecules and other potential drug candidates that have essentially no likelihood of having therapeutic potency high enough to develop into a drug. These rules include

1. # of hydrogen donors ≤ 5 (refers to the total number of hydrogen–nitrogen and hydrogen–oxygen bonds in the small molecule).
2. # of hydrogen acceptors ≤ 10 (refers to the total number of nitrogen and oxygen atoms in the small molecule).
3. A molecular mass of <500 Da.
4. An octanol–water partition coefficient of <5. This value assesses the relationship between lipophilicity and hydrophilicity in order to determine whether the compound is more soluble in fat or water. A coefficient greater than 1 denotes greater fat solubility, while a coefficient lower than 1 denotes greater water solubility.

8.1.5 RNA-SEQ AND HERBAL MEDICINE

Some transcriptome studies have been performed on model plant species such as American ginseng, *Salvia miltiorrhiza* (red sage), and licorice to elucidate novel transcripts. However, many herbs used in traditional medicine are not model organisms and do not have well-characterized genomes or transcriptomes. This lack of substantial genomic information for non-model species makes many aspects of downstream analysis such as analysis of disease pathways, herb cultivation, and drug targets difficult.

Because many medicinal constituents of the TCM methodology are not well-characterized, the de novo assembly tool Trinity is one of the most used tools for RNA-seq analysis because a reference genome is frequently not available. With regard to medicinal plant research, the next section will discuss the role of RNA-seq in mining functional genes in medicinal plants, developing molecular markers,

Herbal Medicine and RNA-seq

exploring the biosynthetic pathways of secondary metabolites, and elucidation of genes involved in stress response for the purposes of cultivation. In addition, applications of RNA sequencing will be discussed as it pertains to molecular mechanisms of disease and identification of new drug targets and new uses for pre-existing drugs.

8.2 MINING FUNCTIONAL GENES OF MEDICINAL PLANTS

Throughout evolution, medicinal plant species have developed regulatory mechanisms that provide them the ability to adapt to fluctuations in external stress stimuli. These regulatory mechanisms promote temporal and geographic distributions in gene expression in many plant species. As a result, typical lab conditions may not be capable of capturing the full scope of the biosynthetic potential of these plants and functionally useful compounds may remain unidentified due to lowly expressed or silent genes. Traditional approaches to identification and purification of NPs and their bioactive constituents are bioactivity-guided techniques. RNA sequencing is advantageous in that we can capture, like a screenshot, of the transcriptomic profile of plant cells in response to purposeful exposure to a particular stress stimulus. Overall, functional gene mining is concerned with identifying biosynthetic pathways, genes that encode key enzymes for production of bioactive compounds, and regulatory mechanisms.

Multiple studies have utilized high-throughput RNA-seq technology paired with cell culture techniques in order to improve our understanding of molecular mechanisms for non-model plant species. One study performed by Xu et al. (2014) used next-generation sequencing to sequence the transcriptome of *Vitis amurensis* (Amur grape) in response to cold exposure. Of the 6,850 unique transcripts attributed to cold regulation, 3,676 transcripts were upregulated, while 3,147 transcripts were downregulated. In addition, transcription factors belonging to 38 distinct families were also implicated in the stress response to cold conditions. A study conducted by Choudhri et al. (2018) utilized Illumina paired-end sequencing to sequence the transcriptomes of root tissue and tissue derived from the leaves of the *Aloe vera* plant. 161,733 unique transcripts were identified from the sequencing of the root tissue, while 221,792 transcripts were sequenced from the leaf-derived tissue samples. In addition, 16 novel genes were identified that were linked to the synthesis of important secondary metabolites including saponins and carotenoids.

A similar study conducted by Singh et al. (2017) used Illumina paired-end sequencing to characterize and sequence the transcriptome of the rhizome *Trillium govanianum*, an herb used for its analgesic and anti-inflammatory properties. In this study, 69,174 unique transcripts were identified, and multiple genes were involved in the synthesis of steroid saponins and other secondary metabolites. Finally, Xu et al. (2016) used next-gen sequencing to characterize the transcriptome of tissues derived from the *Callerya speciosa*. In this study, 4,538 differentially expressed genes (DEGs) were found to be dysregulated that were primarily linked to light responses and synthesis of starch molecules.

212 RNA-seq in Drug Discovery and Development

As we mentioned previously, many herbs and plant species used in traditional medicine are not model organisms with well-characterized genomic profiles. That is, when performing RNA-sequencing experiments, we frequently do not have a reference genome available to annotate our transcripts and thus do not have an idea of their function or to what gene these transcripts map to. An invaluable tool to use in these cases is the Basic Local Alignment Search Tool (BLAST), which can be used to annotate genes across species when a previously unknown gene from one species that is less characterized is compared to a database containing sequences for a more well-characterized organism.

A BLAST query takes as input an amino acid (for proteins) or nucleotide (for DNA/RNA) sequence in FASTA format (from a FASTQ file). The query is then searched against either a specific target sequence or a sequence database containing multiple iterations of our query sequence and our output is a list of sequences with high similarity, usually in the html format. The BLAST program uses a process called seeding in order to identify short matches between our two sequences. When searching for sequence similarities, small sets of letters (words) are used with the default setting in BLAST of $n=3$. For example, if we are performing a protein–protein sequence search using the query sequence YWIMR, the algorithm will search the sequence database for all instances of the words YWI, WIM, and IMR. Once these words have been searched for, the algorithm will compile a matrix composed of all possible neighborhood words, which refers to a word that incorporates a single amino acid substitution (i.e., YQI becomes YAI for instance), as well as a scoring value. By default, BLAST uses the BLOSUM62 scoring scheme which uses a log-odds ratio of the biological probability of the substitution and the random chance probability of the same substitution. Any score greater than the set threshold value T (default=13) is considered significant and will be used in the alignment step of the algorithm. The mapped words will then be extended in both directions, with each extension either decreasing or increasing the associated BLOSUM62 score. Once the alignment phase is complete, any alignments that have a score that exceeds the predetermined threshold value T will be outputted in the BLAST summary report.

To illustrate the use of the BLAST tool, we will perform a simple search using the following nucleotide sequence:

GGTAGCTCTGATAAGAATATTGATGAGTTCATTTCA
AAGCTTGTTTCCTCCTGA.

This sequence was taken from the open-source sequence database Ginseng Genome Database (http://ginsengdb.snu.ac.kr/) and corresponds to a conserved C' terminal domain (last) row in the FASTA file of the sequence of the UDP-glycosyltransferase mRNA in *Panax ginseng* (Korean ginseng). The following steps will be performed using the BLAST tool:

1. Access the BLAST tool using the weblink https://blast.ncbi.nlm.nih.gov/Blast.cgi.
2. Once on the primary user window, click on *Nucleotide BLAST.*

Herbal Medicine and RNA-seq

3. Copy and paste the nucleotide sequence into the *Enter Query Sequence* box.
4. Assign a title to the BLAST search.
5. Under the *Choose Search Set* options, leave the default options as Standard databases and nucleotide collection (nr/nt).
6. Type Korean ginseng in the *Organism* section. Click *add organism* and type Arabidopsis.
7. Under *Program Selection*, select the algorithm Somewhat similar sequences (blastn).
8. Click BLAST at the bottom of the page.

Once we have used the weblink https://blast.ncbi.nlm.nih.gov/Blast.cgi to access the primary user window for the NCBI BLAST tool, we will click on *Nucleotide Blast* (Figure 8.1a). We will then copy our nucleotide sequence GGTAGCTCTGATAAGAATATTGATGAGTTCATTTCAAAGCTTGTTTCCTCCTGA into the *Enter Query Sequence* box (Figure 8.3b). If available, we can instead upload a file containing our sequence query. For this example, we will title the BLAST search Sample nucleotide BLAST (Figure 8.3b). Under the *Choose Search Set* options, we will leave the default options as Standard databases and nucleotide collection (Figure 8.3b). In the *Organism* section, we will first type Korean ginseng. Next, we will click *Add organism* and then type Arabidopsis in the new search bar (Figure 8.3b). In theory, we would be comparing a sequence that corresponds to an unannotated gene from a non-model organism (Korean ginseng) to a well-characterized model organism (*Arabidopsis*). If the sequence is conserved and detected in the model organism, we can then annotate the sequence and know the functionality of the differentially expressed gene/protein. Finally, under *Program Selection*, we will select the somewhat similar

FIGURE 8.3 Overview of the BLAST process. The sequence used for the search query was taken from the open-source sequence database Ginseng Genome Database and corresponds to a conserved sequence of an mRNA of UDP-glycosyltransferase in *Panax ginseng* (Korean ginseng).

sequence (blastn) algorithm and click BLAST at the bottom of the page in order to submit our BLAST search query (Figure 8.3c). Completion of the search query generates a list of sequences that generated significant alignments along with corresponding descriptions, BLOSUM62 score, and accession number (Figure 8.4). As we can see, the sequence is largely conserved between *Panax ginseng* and *Arabidopsis*, further demonstrating that our nucleotide sequence corresponds to the UDP-glycosyltransferase protein.

8.3 DISCOVERY OF SECONDARY METABOLITES AND THEIR METABOLIC PATHWAYS

Secondary metabolites, also referred to as natural products, are organic compounds produced by many different organisms including bacteria and plants that are not involved in the growth and development of the organism. Instead, these secondary metabolites are used to regulate ecological interactions such as deter predation, protect against herbivory, or provide a selective advantage for fecundity and survival. In many instances, these diverse organic metabolic intermediates will have antifungal, antiviral, or antibacterial properties to aid in plant defense. Secondary metabolites are classified based on their chemical and structural composition and can be divided into four primary groups:

1. Terpenes and terpenoids
2. Phenylpropanoids (phenolics)
3. Polyketides
4. Alkaloids

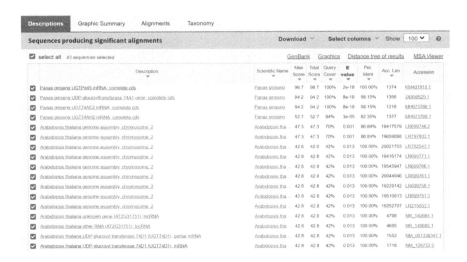

FIGURE 8.4 Sample list of sequences producing significant alignments.

Terpenes comprise the largest group of organic secondary metabolites (60%) and are structurally composed of five-carbon compound units known as isoprenes. Isoprene units that are used as the building blocks for terpenes are derived from isopentenyl pyrophosphate (Figure 8.5a). These hydrocarbons are classified and named based on the number of isoprene units that compose their chemical structure and the chemical formula for such compounds is usually denoted as $(C_5H_8)_n$. Terpenoids are a subclass of terpenes that contain other functional groups in addition to the hydrocarbon isoprene units, particularly oxygen. Terpenoids play a crucial role in the aromatic, flavor, and color properties of many edible plants and spices such as the flavors of cinnamon and clove, the scent of eucalyptus, and the orange color in carrots. Commonly used terpenoids with significant bioactivity include menthol and camphor isolated from the plant *Salvia divinorum*, compounds found in cannabis known as cannabinoids, and the provitamin beta carotene. Saponins are a type of terpenoid that are composed of 30 carbon atoms with at least one glycosylated functional group and are usually toxic organic compounds. They have been shown previously to have anticancer properties as well as a possible treatment for dyslipidemia due to their ability to reduce cholesterol and low-density lipoprotein levels. Figure 8.5b presents the chemical structure for the chemotherapeutic paclitaxel, an anticancer drug classified as a terpenoid.

Phenylpropanoids, also sometimes referred to as the phenolics, are a class of organic compounds characterized by at least one hydroxyl group (-OH) bound directly to an aromatic hydrocarbon ring. Phenylpropanoids are classified based on the number of phenol constituents, which are represented by the chemical formula $(C_6H_5OH)_n$. Tannins are a special class of phenolics that are responsible for the astringent taste in wine, tea, and coffee and are frequently consumed in the human diet from berries, pomegranates, beer, legumes, and nuts. Tannins can also be used to precipitate out proteins, amino acids, and alkaloids, a type of secondary metabolite that will be discussed next. Many drugs have been designed based on isolated phenolic compounds including the pain-reliever acetaminophen (Figure 8.6a), the dopamine prodrug L-DOPA used in the treatment of Parkinson's disease (Figure 8.6b), and the antibiotic amoxicillin (Figure 8.6c).

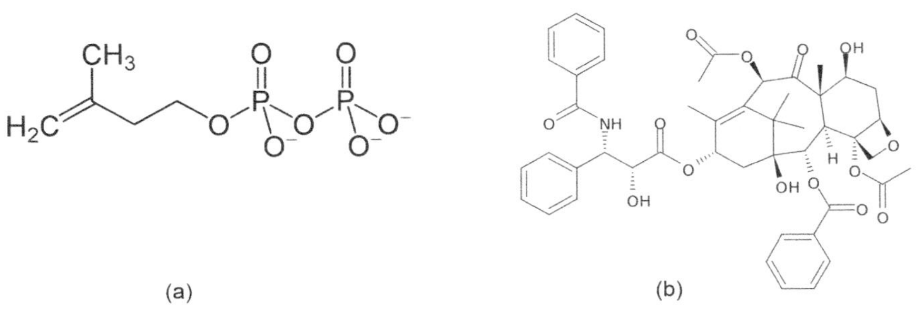

(a) (b)

FIGURE 8.5 The structural formula for isopentenyl pyrophosphate, the source of the hydrocarbons for the isoprene subunits composing terpenes and terpenoids (a), as well as the structure of the anticancer drug paclitaxel (b) are presented.

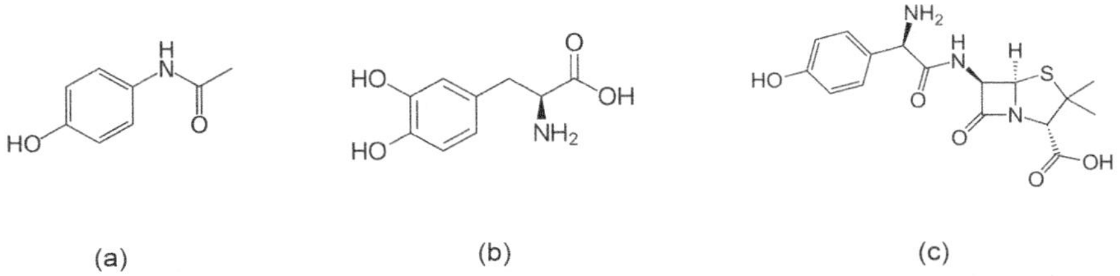

FIGURE 8.6 The structural formulas for the phenolics acetaminophen (a), L-DOPA (b), and amoxicillin (c) are presented.

Alkaloids are the third predominant classification of bioactive secondary metabolites produced by plants and are typically basic organic compounds containing at least one nitrogen atom. Many bioactive compounds that are classified as alkaloids alter components of the central nervous system through disruption of neurotransmitter signaling and are typically bitter/toxic if ingested. Alkaloids can further be classified into either heterocyclic (typical) or non-heterocyclic (atypical). Heterocyclic alkaloids such as nicotine and caffeine have at least two different elements comprising their aromatic rings, typically carbon and another element such as oxygen. In contrast, non-heterocyclic alkaloids such as the gout medication colchicine are composed of aromatic rings with only one elemental species. Many alkaloids with diverse pharmacological profiles have been purified from many different plant species including the anticholinergic drug galantamine, which is used in the treatment of Alzheimer's disease (AD), and the antimalarial drug quinine. Codeine and morphine are alkaloids with analgesic properties and have been isolated from *Papaver somniferum* (opium poppy), while the anticholinergic drug atropine that is used to treat pesticide poisonings and the antitumor drug vincristine have been purified from *Atropa belladonna* (deadly nightshade) and *Catharanthus roseus* (rosy periwinkle), respectively. The chemical structures for the alkaloids nicotine (Figure 8.7a), morphine (Figure 8.7b), and vincristine (Figure 8.7c) are presented.

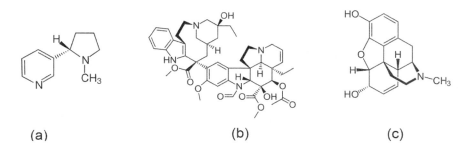

FIGURE 8.7 The structural formulas for the alkaloids nicotine (a), morphine (b), and vincristine (c) are presented.

Herbal Medicine and RNA-seq

Finally, glucosinolates are the fourth significant class of bioactive secondary metabolites produced by many plants and consist of organic compounds that contain both nitrogen and sulfur atoms. Glucosinolates are often synthesized in response to plant damage and are responsible for the pungent smells and taste of such plants as mustard, horseradish, cabbages, and Brussels sprouts. Purified glucosinolates such as allyl isothiocyanate (Figure 8.8a) and glucotropaeolin (Figure 8.8b) are frequently investigated for their possible anticancer and antitumor effects.

Secondary metabolites are essential for a plant's ability to adapt to various developmental and environmental changes. The accumulation and concentration of different bioactive compounds is heavily influenced by environmental factors such as temperature and salinity as well as by developmental stages and tissue localization. Thus, plant samples of the same species may have different distributions of secondary metabolites if they are isolated from plants grown under different conditions or from plants that are currently in different stages of development.

Transcriptomic analysis can be used to identify the biosynthetic pathways that generate bioactive secondary metabolites as well as relevant genes that are triggered in response to different environmental and developmental stimuli. For example, the stems of *Entada phaseoloides* have been used in traditional medicine for years due to its anti-inflammatory and anti-wind-dampness properties, the latter of which refers to the removal of what is termed 'wind' in TCM. Damp wind is usually associated with causing symptoms associated with the common cold including chills, nausea, sore muscles, and listlessness. A study performed by Liao et al. (2020) compared the transcriptomic profiles of tissue samples derived from the root, stem, and leaves of *E. phaseoloides* and found 26 cytochrome p450 as well as 17 uridine diphosphate glycosyltransferase genes associated with the synthesis of triterpenoid saponins, which are the primary bioactive compounds derived from *E. phaseoloides*.

Many other transcriptomic studies have been conducted in order to identify bioactive compounds that may have useful medicinal properties. A comparative transcriptomic analysis was performed by Yan et al. (2020) to assess the differential metabolomic and transcriptomic profiles of extracts taken from purple

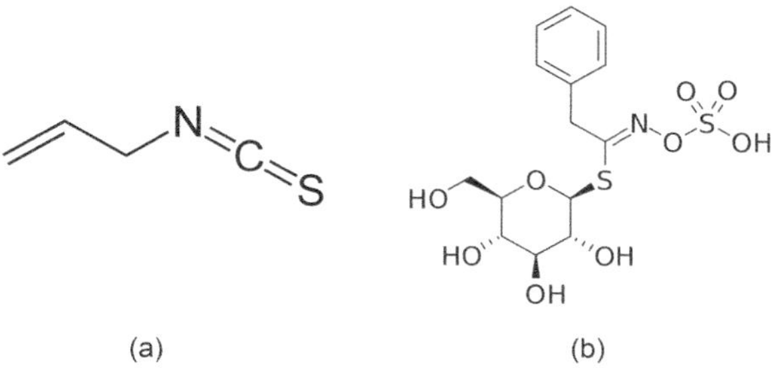

(a)   (b)

FIGURE 8.8   The structural formulas for the glucosinolates allyl isothiocyanate (a) and glucotropaeolin (b) are presented.

and green leaves of the species *Tetrastigma hemsleyanum*. Extracts from *T. hemsleyanum* are commonly used to treat fevers and sore throats due to their broad-spectrum antibiotic effects. In this study, 4,211 unique transcripts and 209 metabolites were found to be differentially expressed with 16 compounds and 14 transcripts demonstrating a strong correlation with the biosynthesis of anthocyanins, compounds in which are derivatives of the phenolic family of secondary metabolites. Finally, the primary bioactive metabolites from *Panax notoginseng*, *Panax notoginseng* saponins (PNS), have been clinically used to treat various cardiovascular diseases. A study conducted by Chen et al. (2021) designed an in vitro myocardial ischemia model in H9C2 cardiomyocytes. The goal of this study was to first subject the cardiomyocytes to oxygen–glucose depletion and then analyze global gene expression in response to treatment with PNS using next-generation sequencing. Thus, they sought to elucidate the mechanisms in which PNS can confer a protective effect on cardiomyocytes. PNS was shown to have anti-apoptotic effects on cardiomyocytes exposed to oxygen–glucose depletion by rescuing cells from cell cycle arrest, inhibiting cardiac hypertrophy, upregulating MAPK and PI3K/AKT signaling pathways, and promoting double-strand break repair mechanisms.

8.4 DISCOVERY OF DEVELOPMENTAL MECHANISMS

Next-generation sequencing and transcriptomics analysis can be used to identify genes involved in the synthesis and regulation of primary metabolites required for the proper growth and development of plants relevant to traditional medicine. In addition, this allows us to assess the differential expression of genes required for plants to respond against abiotic stresses. This provides important insights into more efficient cultivation methods, improved breeding practices, and the facilitation of targeted breeding for the most effective plants.

A comparative study conducted by Rastogi et al. (2019) used next-generation sequencing to observe transcriptome differences in the leaves of *Ocimum tenuiflorum* (holy basil) that are exposed to the following four stress conditions: (1) cold, (2) drought, (3) waterlogged, and (4) excessive salt. Based on statistical significance, 255, 693, 601, and 369 unique transcripts were shown to be upregulated in samples exposed to the abiotic stresses cold, waterlogged, drought, and high salinity, respectively. In addition, 332, 561, 665, and 539 unique transcripts were shown to have decreased expression in leaf samples exposed to the abiotic stresses cold, waterlogged, drought, and high salinity, respectively. 33 of the unique transcripts were shown to be upregulated across all four treatments, 20 of which have been previously linked to stress responses including F-box proteins, cation H(+) antiporter 18-like, and cytosolic sulfotransferase.

Deng et al. (2015) used high-throughput sequencing to elucidate genes required for proper development and growth of seeds derived from *Ephedra przewalskii*, an herb whose primary bioactive ingredients include alkaloids such as ephedrine and pseudoephedrine. This plant has commonly been used to treat mild coughs, pain associated with rheumatoid arthritis (RA), fever, and asthma. A total of

Herbal Medicine and RNA-seq

16,748 unique coding genes were found to be involved in approximately 125 metabolic processes during the process of germination. Findings from this study have helped with the issues of low germination rates, extended growth periods, and overall low survivability of seedlings for this particular plant used in traditional medicine.

Finally, Feng et al. (2020) used RNA-sequencing analysis to compare the full-length transcriptome of wildtype *Angelica sinensis* (female ginseng) with the transcriptome of cultivated *Angelica sinensis* samples. *A. sinensis*, particularly the root of the plant, has been used in traditional medicine to promote healthy blood circulation and serve as an anticoagulant. A total of 25,463 unique transcripts were found to be differentially expressed between the two samples, of which 10,373 transcripts were upregulated in the cultivated *A. sinensis*, while 15,090 transcripts were downregulated when compared to the transcriptome of the wildtype samples. Using Gene Ontology (GO) and Kyoto Encyclopedia of Genes and Genomes (KEGG) enrichment analysis, most transcripts were mapped to MAPK plant–pathogen interactions and plant hormone signal transduction pathways. This knowledge derived from the findings of this study has improved cultivation practices as well as enhanced screening techniques for *A. sinensis*.

8.5 DEVELOPMENT OF MOLECULAR MARKERS TO IMPROVE PLANT BREEDING

Microsatellites, sometimes referred to as simple sequence repeats (SSRs), are small repetitive sequences that are highly prevalent in plant genomes and are more prone to mutations than other regions of the genome. These repeated DNA motifs are usually 1–6 base pairs in length and are repeated between 5 and 50 times within the entirety of the genome. This high degree of genetic diversity (significant polymorphism) in these SSRs is useful in gene mapping, measures of relatedness between subspecies, paternity testing, and selective breeding. A process known as marker-assisted selection is a means to use SSRs as the basis of selective breeding practices. In this method, a trait of interest is identified and selected based on a particular marker, whether morphological or DNA/RNA based, that is linked to the trait of interest such as disease resistance, increased stress tolerance, or greater fecundity. This method assumes that the chosen marker is highly associated with the gene or trait locus of interest due to close proximity and a strong genetic linkage. The use of the marker-assisted selection methodology is appropriate in the following scenarios:

1. The trait of interest manifests in the late stages of plant development. The use of a marker that presents earlier eliminates the need to wait for the plant to become fully developed prior to propagation.
2. The target gene corresponds to a recessive trait.
3. Special conditions are required for the expression of target genes such as when breeding for disease resistance.
4. The phenotype of interest is influenced by two or more unlinked genes.

Multiple published studies have demonstrated the efficacy of using SSRs in plant breeding and the studying of genetic diversity, both within and across species barriers. For example, a study conducted by Kapoor et al. (2020) analyzed the genetic diversity of 48 samples derived from ten different species of asparagus that were grown in various regions across India. The primary bioactive secondary metabolites that are useful in the pharmaceutical industry and traditional medicine include saponins and flavonoids such as rutin and glutathione. 122 allelic variants were amplified using polymerase chain reaction (PCR) and 24 unique polymorphic SSRs were ultimately identified. The findings showed that a significant portion of the genome is conserved across both native and imported species of asparagus. This knowledge allows for greater insight into the genetic composition of asparagus plant species and allows for the identification of more advantageous genotypes for meeting the demands of traditional medicine and drug design.

Another study was performed by Bhandari et al. (2020) which focused on assessing the genetic diversity of *Salvadora oleoides* in arid regions of India. Using Illumina paired-end sequencing, 14,552 unique SSR markers were developed and ultimately 94 primers were successfully amplified, 34 of which were polymorphic in nature. This study was novel in that it was the first study to sequence the transcriptome of this species and is critical in the cultivation and conservation of *Salvadora oleoides*.

Finally, Lade et al. (2020) used RNA-sequencing technology to assess the genetic diversity in 96 samples of *Tinospora cordifolia*, which were taken from ten different regions across India. In this study, 268,149 unique transcripts and 7,611 SSRS were identified with 16 SSRs serving as potential markers of genetic diversity.

8.6   IDENTIFICATION OF TARGET GENES AND MOLECULAR MECHANISMS OF HERBAL DRUGS

In addition to the aforementioned applications of RNA-seq in drug discovery, drug target genes can be identified using common RNA-sequencing methodology. For example, Saini et al. (2017) performed an RNA sequencing-based experiment to elucidate gene signatures responsible for conferring the age-related macular degeneration (AMD) phenotype. AMD is a significant cause of progressive visual impairment and is primarily the result of the accumulation of extracellular debris (drusen) underneath the retinal pigment epithelium, a pigmented monolayer of cells that provides structural support for the neural retina. Using retinal pigment epithelium cells derived from human-induced pluripotent stem cells which expressed the AMD phenotype, this study compared the transcriptomic profiles of the AMD cells with that of wildtype controls. In addition, this study investigated the role of the *ARMS2/HTRA1* genotype in AMD pathology in a subset of the cell population, a genotype which had been previously correlated with increased susceptibility to AMD development. A total of 719 DEGs were identified between the controls and the cells with the *ARMS2/HTRA1* genotype. The study then proceeded to identify possible drugs that could be used to suppress the expression of genes associated with AMD pathology including *CFB*, *APOE*, and *THBS1*. The drug screening process found that nicotinamide (NAM) downregulated 647 of the 719 genes, suggesting that it is an effective means to treat AMD symptoms by suppressing inflammation and production of drusen. Subsequent enrichment analysis of the DEGs found significant NAM-mediated impacts on the macular degeneration, PIK3-Akt, and TGF-α signaling cascades.

A study conducted by Nasrabadi et al. (2019) used the Illumina Hiseq2000 platform to probe the effects of rutin, a NP with antitumor properties found in many fruits and vegetables, on the transcriptome of human colon cancer SW480 cells in order to determine the molecular mechanisms underlying its anticancer effects. Enrichment analysis of DEGs in the cancer cells relative to healthy control cells found that its anticancer effects were linked to alterations in glucose and protein metabolism, regulation of the ER stress response, and induction of apoptotic pathways.

Another study conducted by Wang et al. (2012) sought to understand the mechanistic response of patients with coronary heart disease to Qishenkeli, a TCM herbal mixture containing Radix Astragali Mongolici, salvia miltiorrhiza bunge, Flos Lonicerae, Scrophularia, Radix Aconiti Lateralis Preparata, and Radix Glycyrrhizae. By using a transcriptomic approach, this study found 80 possible new drug candidates, 155 potential novel drug targets, and 279 DEGs.

Two additional studies conducted in recent years further illustrate the efficacy of transcriptomics in elucidating the molecular mechanisms in which herbs and other TCM components elicit their positive physiological responds. Per Lin et al., the molecular mechanisms involved in the antitumor effects of the plant compound shikonin, which is derived from the dried root of *Lithospermum erythrorhizon* commonly referred to as zicao, were explored. Zicao is commonly used as a herbal remedy to reduce inflammation in many East Asian cultures. In this study, three distinct breast cancer cell lines including MCF-7, SK-BR-3, and MDA-MB-231 were utilized to explore the mechanisms in which shikonin promotes cell cycle arrests and induces an apoptotic response. RNA-seq libraries were prepared and sequenced using the Illumina TruSeq RNA Library Prep Kit v2 and Illumina Hiseq2500, respectively, ultimately generating 150-bp paired-end reads. Raw reads were mapped using TopHat and assembled using cufflinks using default parameters. 38 common genes were shown to be differentially expressed across all three cell lines in response to shikonin exposure, which were primarily involved in the MAPK signaling, P53 signaling, antigen processing and presentation, HIF1 signaling, and cell cycle pathways based on KEGG enrichment analysis. The study highlights that shikonin treatment induces the expression of the upstream regulators of MAPK signaling *DUSP1* and *DUSP2*. Upregulation of these two effectors subsequently inhibits downstream JNK and p38 signaling pathways, resulting in cell cycle arrest and apoptosis.

A study conducted by Zhang et al. sought to elucidate the underlying mechanisms triggered by andrographolide, a terpene compound extracted from the herb *Andrographis paniculata*, with regard to modulation of neuroinflammation in AD pathology. Using RNA-seq technology on hippocampus samples derived from the

222 RNA-seq in Drug Discovery and Development

commonly studied APP/PS1 mouse model, several genes including *TLR2*, *CCL3*, *CCL4*, and *TLR1* were found to be downregulated in response to andrographolide exposure. GO and KEGG enrichment analysis showed that the dysregulated transcriptome profile was linked to the alleviation of cognitive impairment, reduced amyloid-β deposition (a hallmark of AD), impaired microglial activity (an immune cell in the brain), and a decrease in the release of pro-inflammatory cytokines. Thus, andrographolide was shown to exhibit neuroprotective effects in an Alzheimer's mouse model.

8.7 SYNERGISM OF HERBAL COMPOUNDS IN PATHWAY REGULATION

A cornerstone of TCM and herbal medicine is the use of multiple plant-based compounds simultaneously in order to elicit a multifaceted physiological response. That is, the use of two or more compounds is used to promote synergistic responses to alleviate particular symptoms such as pain and inflammation. This may include a shared pathway cascade that is enhanced by the introduction of two herbs that each can initiate it in the absence of one of the compounds, albeit with lower efficacy. In contrast, novel regulatory effects may also be unique to the mixture of herbal compounds that are not exhibited when the compounds are used individually. For instance, Qiburi et al. (2021) demonstrated a synergistic effect in modulating stroke-induced neuroinflammation when using a mixture of alantolactone and dehydrodiisoeugenol, which are isolated from *Inula helenium* L. and *Myristica fragrans*, respectively. Model microglial BV2 mouse cells were treated with either alantolactone exclusively, dehydrodiisoeugenol exclusively, or a mixture of the two plant compounds. Raw reads were first processed and then mapped to the mouse reference genome using the HISAT2 pipeline. Differential gene expression analysis was performed using DESeq2 and pathway enrichment analysis was performed using KEGG. 378 DEGs were identified in the alantolactone-treated samples, 957 DEGs were identified in samples treated exclusively with dehydrodiisoeugenol, and 1,718 DEGs were determined from samples treated with the mix of the two compounds. The top five pathways enriched in the samples treated with alantolactone based on KEGG analysis include NOD-like receptor signaling pathway, TNF signaling pathway, cytokine–cytokine receptor interaction, osteoclast differentiation, and cytosolic DNA-sensing pathway. The top five pathways enriched in the samples treated with dehydrodiisoeugenol included TNF signaling pathway, NOD-like receptor signaling pathway, cytokine–cytokine receptor interaction, complement and coagulation cascades, and IL-17 signaling pathway. Finally, the top five pathways enriched in the samples treated with the compound mixture included such pathways as TNF signaling, DNA replication, cell cycle, NOD-like receptor signaling, and steroid biosynthesis. Based on a cutoff criterion of $|\log2FC| \geq 1$, FDR ≤ 0.05, and FPKM ≥ 100, the top 9 downregulated genes including *CCL7*, *CCL2*, *CXCL10*, *Il1β*, and *TNFα* as well as the top 13 upregulated genes are found in the mixed

Herbal Medicine and RNA-seq   223

treatment group. Most of the downregulated genes were linked to inflammation, while many of the upregulated DEGS identified including *SQSTM1*, *PRDX1*, and *CAT* have previously demonstrated antioxidant properties.

A study conducted by Guo et al. (2020) demonstrated the efficacy and synergy of co-administration of imperatorin (IMP), a small molecule isolated from the root of *Angelica dahurica*, and β-sitosterol (STO), a plant sterol similar in composition to cholesterol that is extracted from many different plants, in the treatment of collagen-induced RA. RNA sequencing was performed on isolated CD4$^+$ T cells derived from peripheral blood samples taken from healthy rats, the CIA model rats presenting with collagen-induced RA, model rats treated with IMP only, model rats treated with β-sitosterol only, and model rates treated with both plant compounds. When compared to the healthy controls, 347 DEGs were identified in the untreated CIA model samples. The transcriptomic profiles of the untreated CIA models were then compared to the IMP-treated group (183 DEGs), the β-sitosterol-treated group (243 DEGs), and the group treated with both IMP and β-sitosterol (202 DEGs). The top five canonical pathways found to be dysregulated in the synergistic treatment group were altered T- and B-cell signaling in RA, TREM1 signaling, role of Mφs macrophages, fibroblasts, and endothelial cells in RA, role of osteoblasts, osteoclasts, and chondrocytes in RA, and the unfolded protein response. In particular, the genes *LTA, CD83,* and *SREBF1* were found to be significantly upregulated in the IMP-STO treatment group yet significantly downregulated in the untreated model group. Finally, they compared the DEGs that were differentially modulated between the IMP-STO treatment group and the single treatment groups for each of the five major dysregulated canonical pathways. For instance, the genes *NF-κB, NF-κB2, LTA,* and *RELB* were found to have decreased expression in the untreated model group. Treatment with either IMP or STO individually resulted in the downregulation of *NF-κB* and *NF-κB2*. However, treatment with both IMP and STO resulted in the downregulation of the pro-inflammatory *IL-1* while simultaneously downregulating *LTA* and *TGF-β*. Thus, this study showed differences in how transcriptome profiles are regulated between cells treated with either IMP or STO and cells that are concomitantly treated with both natural compounds.

8.8   HERBAL MEDICINE TOXICITY

Many individuals, both in Western and Eastern countries, are exposed to various herbal remedies and supplements, many of which consist of TCM and other components of traditional herbal medicine. Although many herbs in traditional medicine are effective at low doses and are used in conjunction with multiple herbs to confer an effective treatment, toxicity profiles of many components are not well characterized, and supplements may expose individuals to potentially toxic doses of a given herb. Because most supplements are taken orally, many toxicity studies involving microarray and RNA sequencing focus on transcriptomic changes in the liver in order to assess the toxicity of an herb or supplement.

The liver is primarily responsible for the detoxification and metabolism of orally administered compounds and thus is a prime target for toxicity-induced injury. A study conducted by Tu et al. (2021) utilized a combined transcriptomics/metabolomics/proteomics approach to elucidate the molecular mechanism of hepatotoxicity for the commonly used TCM Dictamni Cortex extract, which is purified from the dried root of *Dictamnus dasycarpus Turcz.* HepaRG cell samples were divided into three distinct groups including control (untreated), dictamnine treated-1-passage (P1) group, and dictamnine treated-5-passage (P5) group. When comparing the P1 treatment group with the untreated control group, 33 DEGs were identified with 19 exhibiting enhanced expression and 14 genes being downregulated. 207 DEGs were identified in the P5 vs. untreated control comparison with 106 upregulated and 101 downregulated genes. In addition, female mice were shown to be more susceptible to greater hepatotoxicity in a dose-dependent manner. The pooled DEGs from both experimental comparisons were then pooled together and KEGG and GO analyses were performed. Significant pathways dysregulated by dictamnine treatment included oxidative stress, hepatic cell apoptosis, lipid synthesis, and lipid metabolism. Increased oxidative stress and hepatic cell death were due to the downregulation of many antioxidative enzymes including *CAT*, *SOD*, and GP_x-1 as well as the upregulation of *GSTA1* and the proapoptotic factor *BAX*. In addition, heightened disordered lipid metabolism was induced by the downregulation of *ACAT1* and *FABP-1* as well as the subsequent upregulation of *ACSL4*. Thus, this study determined the molecular mechanism in which the natural compound dictamnine induced hepatotoxicity.

In addition to detecting how a given herbal compound can promote hepatotoxicity, transcriptomics can also be used to identify how herbal compounds can infer a protective effect against cytotoxicity. For instance, a study conducted by Chen et al. (2022) used a transcriptomics approach to determine the mechanisms in which ethanol extract of saffron (ESS), which is taken from the saffron plant *Crocus sativus* and is traditionally used to treat mental disorders such as anxiety, confers a protective effect of PC12 cells in the brain against cytotoxicity induced by corticosterone exposure. Prepped libraries were sequenced using the Illumina Hiseq4000 platform and paired-end reads of length 150 bp were generated. DESeq2 was used to assess DEGs and functional enrichment analysis was performed using both GO terms and KEGG pathway analysis. When comparing the transcriptomic profiles of the negative control group and the model PC12 cells treated with 500 µM of corticosterone for 24 hours, 998 DEGs were identified, 413 of which were upregulated and 585 of which were downregulated. 246 DEGs (137 upregulated and 109 downregulated) were identified when comparing corticosterone-treated PC12 cells and PC12 cells treated with both corticosterone and ESS. The 138 genes that were differentially expressed between the model group and the group treated with ESS, which included 45 DEGs upregulated and 93 DEGs downregulated after treatment with ESS, were then subjected to GO and KEGG analysis. The top ten pathways enriched by ESS treatment based on KEGG analysis included ferroptosis, endometrial cancer, p53 signaling, protein processing in endoplasmic reticulum, thyroid cancer, hepatocellular carcinoma,

Herbal Medicine and RNA-seq 225

transcriptional regulation in cancer, apoptosis, and MAPK signaling with the most genes mapping to the MAPK signaling pathway.

A study conducted by Liu et al. (2022) utilized the principles of RNA sequencing to assess the mechanisms in which the TCM Hua-Feng-Dan (HFD) induces an adaptive response in the liver. Hua-Feng-Dan is an herbal remedy that has been used to treat stroke and other medical conditions for hundreds of years. Hua-Feng-Dan consists of a variety of minerals such as cinnabar and realgar as well as multiple herbal species including *Gastrodia elata Blume*, *Nepeta tenuifolia Benth*, *Croton tiglium L.*, *Atractylodes lancea*, and *Typhonium giganteum Engl.*, among others. Using network pharmacology and predictive techniques, β-sitosterol, luteolin, baicalein, and wogonin were identified as possible significant active constituents of Hua-Feng-Dan. In addition, Hua-Feng-Dan is usually administered alongside a fermented product referred to as Yaomu. Yaomu is considered as a guide drug in TCM, a term in which refers to a compound that is added to a traditional medicine regiment in order to increase efficacy and reduce toxicity of the treatment. Adult male C57BL/6J mice were randomly divided into four groups including the control group ($n = 5$), YM-0.1 ($n = 5$, Yaomu 0.1 g/kg), YM-0.3 ($n = 7$, Yaomu 0.3 g/kg), and HFD ($n = 5$, Hua-Feng-Dan 1.2 g/kg). After 7 days of treatment and a wait period of 24 hours, the mice were anesthetized, and livers were collected for sequencing. DESeq2 methodology was used to identify DEGs ($p < 0.05$) between untreated control cells and each of the treatment groups. The top seven enriched KEGG pathways identified for the comparison of untreated control cells and the HFD group included cholesterol metabolism, bile secretion, PPAR signaling, drug metabolism, fat digestion and absorption, and retinol metabolism (total DEGs = 806). In addition, 235 DEGs were identified from the control vs. YM-0.1 comparison and 92 DEGs were determined for the control vs. YM-0.3 comparison. Two-dimensional clustering was then used to visualize the differences in gene expression between the three groups. The following four clusters were selected for further annotation:

1. Upregulated cluster lines 42–150 (DEGs = 109 genes), which were increased in only the HFD and YM-0.1 groups and were linked to cellular function and signal regulation pathways.
2. Upregulated cluster lines 151–179 (DEGs = 28 genes), which were increased in all three treatment groups and were linked to cellular function, circadian rhythm, and signal transduction pathways.
3. Downregulated cluster lines 494–556 (DEGs = 63 genes), which were decreased in only the HFD and YM-0.1 groups and were linked to metabolism and immunomodulation pathways.
4. Downregulated cluster lines 557–567 (DEGs = 11 genes), which were decreased in all three treatment groups and were linked to metabolism and immunomodulation pathways.

Thus, this study elucidated the molecular mechanisms and underlying transcriptomic profile changes of how Hua-Feng-Dan initiates adaptive responses in the liver without resulting in hepatotoxicity.

8.9 NATURAL DRUG REPURPOSING

Drug repurposing, otherwise referred to as drug repositioning, is defined as using already existing and approved therapeutics for new disease targets. As drug discovery is a very time-consuming and costly process, drug repurposing is an effective and faster approach to disease treatment. With drug repurposing, we can effectively skip phase 1 trials as the safety and pharmacokinetics for the pre-existing compound is already known. In addition, because the production and supply chain logistics have already been implemented, production can be more easily scaled up for increased production and release of the drug. However, one possible disadvantage is if the dosage required for effective treatment differs between the initial condition the drug was designed for and the possible new target due to drug repositioning. If the effective dosage differs, the phase 1 trials cannot be safely skipped and thus would not offer much advantages over the traditional drug discovery pipeline.

Several approaches may be used to find new disease pathways that can be treated with previously designed drugs. The first method involves structural modeling and docking methods to identify new drug targets based on structural compatibility. In contrast, transcriptomic approaches are more concerned with identifying diseases with similar underlying mechanisms responsible for their pathology. This is done by comparing a list of DEGs for one drug to the perturbation profile of another drug and looking for overlap between the two lists. For example, a study conducted by Bacelli et al. (2019) screened for possible therapeutic targets in samples presenting with adult acute myeloid leukemia (AML) using mubritinib, a protein kinase inhibitor that was originally designed for breast cancer treatment. AML cases can be classified into 20 different variants based upon various genetic signatures. This study concluded that AML cases that are chemotherapy-sensitive yet show elevated levels of hypoxia markers were more resistant to mubritinib. However, mubritinib was particularly effective in promoting apoptosis in ERBB2$^+$-positive AML cases through inhibition of oxidative phosphorylation pathways.

Namikawa et al. (2019) performed a transcriptomic study focused on the development of antivirulence therapeutics for the treatment of hypervirulent *Klebsiella pneumoniae* infection. The antibiotic rifampicin was repurposed and found to have significant anti-mucoviscous activity by regulating the expression of genes involved in capsular polysaccharide synthesis as well as downregulation of the regulator of mucoid phenotype A.

Finally, a transcriptome-guided approach was utilized by Arakelyan et al. (2019) to repurpose infliximab, a drug previously used in the treatment of ulcerative colitis and Crohn's disease, into a possible therapeutic for sarcoidosis.

8.10 SUMMARY

In this chapter, we introduced the idea of traditional medicine as well as identified modern virtual screening methods such as molecular docking and

Herbal Medicine and RNA-seq

pharmacophore modeling. We discussed the idea of network pharmacology as well as the important components of constructing interaction networks including databases such as KEGG and DrugBank in which we can obtain information on chemical compounds, pretreatment of compounds to eliminate compounds with little drug potential, and various virtual screening approaches. Finally, this chapter discussed applications of transcriptomic studies in the drug discovery process including functional gene mining, discovery of secondary metabolites, elucidation of developmental and molecular mechanisms, and determination of toxicity profiles of traditional herbal medicines.

KEYWORDS AND PHRASES

After reading this chapter, you should be able to demonstrate familiarity with the following words and phrases:

- Understand and be able to define TCM.
- Understand and be able to describe the four major steps of network pharmacology.
- Be able to identify public databases in which drug information can be obtained from.
- Understand how transcriptomics can be used in various applications including functional gene mining, discovering novel secondary metabolites, elucidating developmental mechanisms, identifying biomarkers for improved plant breeding, determining mechanisms of action for herbal treatments, determining how different herbal components work in a synergistic manner, determining the toxicity profile of TCM components, and repurposing pre-existing NPs for treatment of different diseases.

BIBLIOGRAPHY

Arakelyan, A., Nersisyan, L., Nikoghosyan, M., Hakobyan, S., Simonyan, A., Hopp, L., Loeffler-Wirth, H., & Binder, H. (2019, Dec 12). Transcriptome-guided drug repositioning. *Pharmaceutics, 11*(12), 677. https://doi.org/10.3390/pharmaceutics11120677

Atanasov, A. G., Zotchev, S. B., Dirsch, V. M., & Supuran, C. T. (2021, Mar). Natural products in drug discovery: Advances and opportunities. *Nat Rev Drug Discov, 20*(3), 200–216. https://doi.org/10.1038/s41573-020-00114-z

Baccelli, I., Gareau, Y., Lehnertz, B., Gingras, S., Spinella, J. F., Corneau, S., Mayotte, N., Girard, S., Frechette, M., Blouin-Chagnon, V., Leveille, K., Boivin, I., MacRae, T., Krosl, J., Thiollier, C., Lavallee, V. P., Kanshin, E., Bertomeu, T., Coulombe-Huntington, J., St-Denis, C., Bordeleau, M. E., Boucher, G., Roux, P. P., Lemieux, S., Tyers, M., Thibault, P., Hebert, J., Marinier, A., & Sauvageau, G. (2019, Jul 8). Mubritinib targets the electron transport chain complex I and reveals the landscape of OXPHOS dependency in acute myeloid leukemia. *Cancer Cell, 36*(1), 84–99 e88. https://doi.org/10.1016/j.ccell.2019.06.003

Becker, K. G., Barnes, K. C., Bright, T. J., & Wang, S. A. (2004, May). The genetic association database. *Nat Genet, 36*(5), 431–432. https://doi.org/10.1038/ng0504-431

Bhandari, M. S., Meena, R. K., Shamoon, A., Saroj, S., Kant, R., & Pandey, S. (2020, Sep). First de novo genome specific development, characterization and validation of simple sequence repeat (SSR) markers in Genus Salvadora. *Mol Biol Rep, 47*(9), 6997–7008. https://doi.org/10.1007/s11033-020-05758-z

Chen, C., Wang, T., Wu, F., Huang, W., He, G., Ouyang, L., Xiang, M., Peng, C., & Jiang, Q. (2014). Combining structure-based pharmacophore modeling, virtual screening, and in silico ADMET analysis to discover novel tetrahydro-quinoline based pyruvate kinase isozyme M2 activators with antitumor activity. *Drug Des Devel Ther, 8*, 1195–1210. https://doi.org/10.2147/DDDT.S62921

Chen, C. Y. (2011, Jan 6). TCM Database@Taiwan: The world's largest traditional Chinese medicine database for drug screening in silico. *PLoS One, 6*(1), e15939. https://doi.org/10.1371/journal.pone.0015939

Chen, L., Zhang, Y. H., Zheng, M., Huang, T., & Cai, Y. D. (2016, Dec). Identification of compound-protein interactions through the analysis of gene ontology, KEGG enrichment for proteins and molecular fragments of compounds. *Mol Genet Genomics, 291*(6), 2065–2079. https://doi.org/10.1007/s00438-016-1240-x

Chen, S., Wu, Y., Qin, X., Wen, P., Liu, J., & Yang, M. (2021, Mar 25). Global gene expression analysis using RNA-seq reveals the new roles of *Panax notoginseng* Saponins in ischemic cardiomyocytes. *J Ethnopharmacol, 268*, 113639. https://doi.org/10.1016/j.jep.2020.113639

Chen, X., Yang, T., Zhang, C., & Ma, Z. (2022, Jan 31). RNA-seq based transcriptome analysis of ethanol extract of saffron protective effect against corticosterone-induced PC12 cell injury. *BMC Complement Med Ther, 22*(1), 29. https://doi.org/10.1186/s12906-022-03516-1

Choudhri, P., Rani, M., Sangwan, R. S., Kumar, R., Kumar, A., & Chhokar, V. (2018, Jun 1). De novo sequencing, assembly and characterisation of Aloe vera transcriptome and analysis of expression profiles of genes related to saponin and anthraquinone metabolism. *BMC Genomics, 19*(1), 427. https://doi.org/10.1186/s12864-018-4819-2

Dekkers, J. C., & Hospital, F. (2002, Jan). The use of molecular genetics in the improvement of agricultural populations. *Nat Rev Genet, 3*(1), 22–32. https://doi.org/10.1038/nrg701

Deng N., Shi S. Q., Chang E. M., Liu J. F., Lan Q., & Jiang Z. P. (2015). Transcriptomic analysis of germinated seeds of *Ephedra przewalskii*. *J Northeast For Univ, 43*(2),28–32.

Ejelonu, O. C., Elekofehinti, O. O., & Adanlawo, I. G. (2017, Mar). Tithonia diversifolia saponin-blood lipid interaction and its influence on immune system of normal wistar rats. *Biomed Pharmacother, 87*, 589–595. https://doi.org/10.1016/j.biopha.2017.01.017

Feng, W. M., Liu, P., Yan, H., Yu, G., Guo, Z. X., Zhu, L., Ma, J. W., Qian, D. W., & Duan, J. A. (2020, Apr). [Transcriptomic data analyses of wild and cultivated Angelica sinensis root by high-throughput sequencing technology]. *Zhongguo Zhong Yao Za Zhi, 45*(8), 1879–1886. https://doi.org/10.19540/j.cnki.cjcmm.20200208.101

Fokunang, C. N., Ndikum, V., Tabi, O. Y., Jiofack, R. B., Ngameni, B., Guedje, N. M., Tembe-Fokunang, E. A., Tomkins, P., Barkwan, S., Kechia, F., Asongalem, E., Ngoupayou, J., Torimiro, N. J., Gonsu, K. H., Sielinou, V., Ngadjui, B. T., Angwafor, F., 3rd, Nkongmeneck, A., Abena, O. M., Ngogang, J., Asonganyi, T., Colizzi, V., Lohoue, J., & Kamsu, K. (2011). Traditional medicine: past, present and future research and development prospects and integration in the National Health System of Cameroon. *Afr J Tradit Complement Altern Med, 8*(3), 284–295. https://doi.org/10.4314/ajtcam.v8i3.65276

Fu, Y., Luo, J., Qin, J., & Yang, M. (2019, May 10). Screening techniques for the identification of bioactive compounds in natural products. *J Pharm Biomed Anal, 168*, 189–200. https://doi.org/10.1016/j.jpba.2019.02.027

Guo, Q., Li, L., Zheng, K., Zheng, G., Shu, H., Shi, Y., Lu, C., Shu, J., Guan, D., Lu, A., & He, X. (2020, Aug). Imperatorin and beta-sitosterol have synergistic activities in alleviating collagen-induced arthritis. *J Leukoc Biol, 108*(2), 509–517. https://doi.org/10.1002/JLB.3MA0320-440RR

Hamosh, A., Scott, A. F., Amberger, J., Bocchini, C., Valle, D., & McKusick, V. A. (2002, Jan 1). Online Mendelian Inheritance in Man (OMIM), a knowledgebase of human genes and genetic disorders. *Nucleic Acids Res, 30*(1), 52–55. https://doi.org/10.1093/nar/30.1.52

Hao, D. C., & Xiao, P. G. (2015). Genomics and evolution in traditional medicinal plants: road to a healthier life. *Evol Bioinform Online, 11*, 197–212. https://doi.org/10.4137/EBO.S31326

Hong, M., Wang, X. Z., Wang, L., Hua, Y. Q., Wen, H. M., & Duan, J. A. (2011, Jan 5). Screening of immunomodulatory components in Yu-ping-feng-san using splenocyte binding and HPLC. *J Pharm Biomed Anal, 54*(1), 87–93. https://doi.org/10.1016/j.jpba.2010.08.016

Hopkins, A. L. (2008, Nov). Network pharmacology: the next paradigm in drug discovery. *Nat Chem Biol, 4*(11), 682–690. https://doi.org/10.1038/nchembio.118

Kapoor, M., Mawal, P., Sharma, V., & Gupta, R. C. (2020, Sep 14). Analysis of genetic diversity and population structure in Asparagus species using SSR markers. *J Genet Eng Biotechnol, 18*(1), 50. https://doi.org/10.1186/s43141-020-00065-3

Kumar, A., & Zhang, K. Y. J. (2018). Advances in the development of shape similarity methods and their application in drug discovery. *Front Chem, 6*, 315. https://doi.org/10.3389/fchem.2018.00315

Lade, S., Pande, V., Rana, T. S., & Yadav, H. K. (2020, Jul). Estimation of genetic diversity and population structure in *Tinospora cordifolia* using SSR markers. *3 Biotech, 10*(7), 310. https://doi.org/10.1007/s13205-020-02300-7

Liao, W., Mei, Z., Miao, L., Liu, P., & Gao, R. (2020, Sep 15). Comparative transcriptome analysis of root, stem, and leaf tissues of *Entada phaseoloides* reveals potential genes involved in triterpenoid saponin biosynthesis. *BMC Genomics, 21*(1), 639. https://doi.org/10.1186/s12864-020-07056-1

Lin, K. H., Huang, M. Y., Cheng, W. C., Wang, S. C., Fang, S. H., Tu, H. P., Su, C. C., Hung, Y. L., Liu, P. L., Chen, C. S., Wang, Y. T., & Li, C. Y. (2018, Feb 8). RNA-seq transcriptome analysis of breast cancer cell lines under shikonin treatment. *Sci Rep, 8*(1), 2672. https://doi.org/10.1038/s41598-018-21065-x

Liu, J. J., Liang, Y., Zhang, Y., Wu, R. X., Song, Y. L., Zhang, F., Shi, J. S., Liu, J., Xu, S. F., & Wang, Z. (2022). GC-MS profile of Hua-Feng-Dan and RNA-seq analysis of induced adaptive responses in the liver. *Front Pharmacol, 13*, 730318. https://doi.org/10.3389/fphar.2022.730318

Lv, C., Wu, X., Wang, X., Su, J., Zeng, H., Zhao, J., Lin, S., Liu, R., Li, H., Li, X., & Zhang, W. (2017, Mar 23). The gene expression profiles in response to 102 traditional Chinese medicine (TCM) components: A general template for research on TCMs. *Sci Rep, 7*(1), 352. https://doi.org/10.1038/s41598-017-00535-8

Matthews, L., Gopinath, G., Gillespie, M., Caudy, M., Croft, D., de Bono, B., Garapati, P., Hemish, J., Hermjakob, H., Jassal, B., Kanapin, A., Lewis, S., Mahajan, S., May, B., Schmidt, E., Vastrik, I., Wu, G., Birney, E., Stein, L., & D'Eustachio, P. (2009, Jan). Reactome knowledgebase of human biological pathways and processes. *Nucleic Acids Res, 37*, D619–622. https://doi.org/10.1093/nar/gkn863

McGinnis, S., & Madden, T. L. (2004, Jul 1). BLAST: At the core of a powerful and diverse set of sequence analysis tools. *Nucleic Acids Res, 32*, W20–25. https://doi.org/10.1093/nar/gkh435

Meng, X. Y., Zhang, H. X., Mezei, M., & Cui, M. (2011, Jun). Molecular docking: A powerful approach for structure-based drug discovery. *Curr Comput Aided Drug Des, 7*(2), 146–157. https://doi.org/10.2174/157340911795677602

Moses, T., Papadopoulou, K. K., & Osbourn, A. (2014, Nov-Dec). Metabolic and functional diversity of saponins, biosynthetic intermediates and semi-synthetic derivatives. *Crit Rev Biochem Mol Biol, 49*(6), 439–462. https://doi.org/10.3109/10409238.2014.953628

Muhammad, J., Khan, A., Ali, A., Fang, L., Yanjing, W., Xu, Q., & Wei, D. Q. (2018). Network pharmacology: Exploring the resources and methodologies. *Curr Top Med Chem, 18*(12), 949–964. https://doi.org/10.2174/1568026618666180330141351

Namikawa, H., Oinuma, K. I., Sakiyama, A., Tsubouchi, T., Tahara, Y. O., Yamada, K., Niki, M., Takemoto, Y., Miyata, M., Kaneko, Y., Shuto, T., & Kakeya, H. (2019, Aug). Discovery of anti-mucoviscous activity of rifampicin and its potential as a candidate antivirulence agent against hypervirulent *Klebsiella pneumoniae*. *Int J Antimicrob Agents, 54*(2), 167–175. https://doi.org/10.1016/j.ijantimicag.2019.05.018

Nasri Nasrabadi, P., Zareian, S., Nayeri, Z., Salmanipour, R., Parsafar, S., Gharib, E., Asadzadeh Aghdaei, H., & Zali, M. R. (2019, Sep). A detailed image of rutin underlying intracellular signaling pathways in human SW480 colorectal cancer cells based on miRNAs-lncRNAs-mRNAs-TFs interactions. *J Cell Physiol, 234*(9), 15570–15580. https://doi.org/10.1002/jcp.28204

Navarova, H., Bernsdorff, F., Doring, A. C., & Zeier, J. (2012, Dec). Pipecolic acid, an endogenous mediator of defense amplification and priming, is a critical regulator of inducible plant immunity. *Plant Cell, 24*(12), 5123–5141. https://doi.org/10.1105/tpc.112.103564

Qian, Z. M., Qin, S. J., Yi, L., Li, H. J., Li, P., & Wen, X. D. (2008, Feb). Binding study of Flos Lonicerae Japonicae with bovine serum albumin using centrifugal ultrafiltration and liquid chromatography. *Biomed Chromatogr, 22*(2), 202–206. https://doi.org/10.1002/bmc.916

Qiburi, Q., Temuqile, T., & Baigude, H. (2021). Synergistic regulation of microglia gene expression by natural molecules in herbal medicine. *Evid Based Complement Alternat Med, 2021*, 9920364. https://doi.org/10.1155/2021/9920364

Qiu, J. Y., Chen, X., Zheng, X. X., Jiang, X. L., Yang, D. Z., Yu, Y. Y., Du, Q., Tang, D. Q., & Yin, X. X. (2015, Feb). Target cell extraction coupled with LC-MS/MS analysis for screening potential bioactive components in Ginkgo biloba extract with preventive effect against diabetic nephropathy. *Biomed Chromatogr, 29*(2), 226–232. https://doi.org/10.1002/bmc.3264

Rastogi, S., Shah, S., Kumar, R., Vashisth, D., Akhtar, M. Q., Kumar, A., Dwivedi, U. N., & Shasany, A. K. (2019). Ocimum metabolomics in response to abiotic stresses: Cold, flood, drought and salinity. *PLoS One, 14*(2), e0210903. https://doi.org/10.1371/journal.pone.0210903

Saini, J. S., Corneo, B., Miller, J. D., Kiehl, T. R., Wang, Q., Boles, N. C., Blenkinsop, T. A., Stern, J. H., & Temple, S. (2017, May 4). Nicotinamide ameliorates disease phenotypes in a human ipsc model of age-related macular degeneration. *Cell Stem Cell, 20*(5), 635–647. https://doi.org/10.1016/j.stem.2016.12.015

Scherlach, K., & Hertweck, C. (2021, Jun 23). Mining and unearthing hidden biosynthetic potential. *Nat Commun, 12*(1), 3864. https://doi.org/10.1038/s41467-021-24133-5

Shannon, P., Markiel, A., Ozier, O., Baliga, N. S., Wang, J. T., Ramage, D., Amin, N., Schwikowski, B., & Ideker, T. (2003, Nov). Cytoscape: A software environment for integrated models of biomolecular interaction networks. *Genome Res, 13*(11), 2498–2504. https://doi.org/10.1101/gr.1239303

Singh, P., Singh, G., Bhandawat, A., Singh, G., Parmar, R., Seth, R., & Sharma, R. K. (2017, Mar 28). Spatial transcriptome analysis provides insights of key gene(s) involved in steroidal Saponin biosynthesis in medicinally important herb *Trillium govanianum. Sci Rep, 7*, 45295. https://doi.org/10.1038/srep45295

Su, S. L., Yu, L., Hua, Y. Q., Duan, J. A., Deng, H. S., Tang, Y. P., Lu, Y., & Ding, A. W. (2008, Dec). Screening and analyzing the potential bioactive components from Shaofu Zhuyu decoction, using human umbilical vein endothelial cell extraction and high-performance liquid chromatography coupled with mass spectrometry. *Biomed Chromatogr, 22*(12), 1385–1392. https://doi.org/10.1002/bmc.1070

Sun, M., Huang, L., Zhu, J., Bu, W., Sun, J., & Fang, Z. (2015, Jun). Screening nephroprotective compounds from cortex Moutan by mesangial cell extraction and UPLC. *Arch Pharm Res, 38*(6), 1044–1053. https://doi.org/10.1007/s12272-014-0469-3

Sun, M., Ren, J., Du, H., Zhang, Y., Zhang, J., Wang, S., & He, L. (2010, Oct 15). A combined A431 cell membrane chromatography and online high performance liquid chromatography/mass spectrometry method for screening compounds from total alkaloid of Radix Caulophylli acting on the human EGFR. *J Chromatogr B Analyt Technol Biomed Life Sci, 878*(28), 2712–2718. https://doi.org/10.1016/j.jchromb.2010.08.010

Szklarczyk, D., Franceschini, A., Wyder, S., Forslund, K., Heller, D., Huerta-Cepas, J., Simonovic, M., Roth, A., Santos, A., Tsafou, K. P., Kuhn, M., Bork, P., Jensen, L. J., & von Mering, C. (2015, Jan). String v10: Protein-protein interaction networks, integrated over the tree of life. *Nucleic Acids Res, 43*, D447–452. https://doi.org/10.1093/nar/gku1003

Tu, C., Xu, Z., Tian, L., Yu, Z., Wang, T., Guo, Z., Zhang, J., & Wang, T. (2021). Multiomics integration to reveal the mechanism of hepatotoxicity induced by dictamnine. *Front Cell Dev Biol, 9*, 700120. https://doi.org/10.3389/fcell.2021.700120

Wang, C., Hu, S., Chen, X., & Bai, X. (2016, May). Screening and quantification of anticancer compounds in traditional Chinese medicine by hollow fiber cell fishing and hollow fiber liquid/solid-phase microextraction. *J Sep Sci, 39*(10), 1814–1824. https://doi.org/10.1002/jssc.201600103

Wang, J., & Li, X. J. (2011, Aug). [Network pharmacology and drug discovery]. *Sheng Li Ke Xue Jin Zhan, 42*(4), 241–245. https://www.ncbi.nlm.nih.gov/pubmed/22066413

Wang, Y., Liu, Z., Li, C., Li, D., Ouyang, Y., Yu, J., Guo, S., He, F., & Wang, W. (2012). Drug target prediction based on the herbs components: The study on the multitargets pharmacological mechanism of Qishenkeli acting on the coronary heart disease. *Evid Based Complement Alternat Med, 2012*, 698531. https://doi.org/10.1155/2012/698531

Wang, Y., Xiao, J., Suzek, T. O., Zhang, J., Wang, J., & Bryant, S. H. (2009, Jul). PubChem: A public information system for analyzing bioactivities of small molecules. *Nucleic Acids Res, 37*, W623–633. https://doi.org/10.1093/nar/gkp456

Wishart, D. S., Knox, C., Guo, A. C., Shrivastava, S., Hassanali, M., Stothard, P., Chang, Z., & Woolsey, J. (2006, Jan 1). DrugBank: A comprehensive resource for in silico drug discovery and exploration. *Nucleic Acids Res, 34*, D668–D672. https://doi.org/10.1093/nar/gkj067

Xin, J., Zhang, R. C., Wang, L., & Zhang, Y. Q. (2017). Researches on transcriptome sequencing in the study of traditional Chinese medicine. *Evid Based Complement Alternat Med, 2017*, 7521363. https://doi.org/10.1155/2017/7521363

Xu, L., Wang, J., Lei, M., Li, L., Fu, Y., Wang, Z., Ao, M., & Li, Z. (2016). Transcriptome analysis of storage roots and fibrous roots of the traditional medicinal herb *Callerya speciosa* (Champ.) *ScHot. PLoS One, 11*(8), e0160338. https://doi.org/10.1371/journal.pone.0160338

Xu, W., Li, R., Zhang, N., Ma, F., Jiao, Y., & Wang, Z. (2014, Nov). Transcriptome profiling of *Vitis amurensis*, an extremely cold-tolerant Chinese wild Vitis species, reveals candidate genes and events that potentially connected to cold stress. *Plant Mol Biol, 86*(4–5), 527–541. https://doi.org/10.1007/s11103-014-0245-2

Yan, J., Qian, L., Zhu, W., Qiu, J., Lu, Q., Wang, X., Wu, Q., Ruan, S., & Huang, Y. (2020). Integrated analysis of the transcriptome and metabolome of purple and green leaves of Tetrastigma hemsleyanum reveals gene expression patterns involved in anthocyanin biosynthesis. *PLoS One, 15*(3), e0230154. https://doi.org/10.1371/journal.pone.0230154

Yang, X., Kui, L., Tang, M., Li, D., Wei, K., Chen, W., Miao, J., & Dong, Y. (2020). High-throughput transcriptome profiling in drug and biomarker discovery. *Front Genet, 11*, 19. https://doi.org/10.3389/fgene.2020.00019

Yi, F., Li, L., Xu, L. J., Meng, H., Dong, Y. M., Liu, H. B., & Xiao, P. G. (2018). In silico approach in reveal traditional medicine plants pharmacological material basis. *Chin Med, 13*, 33. https://doi.org/10.1186/s13020-018-0190-0

Zhang, Y., He, L., Meng, L., & Luo, W. (2008, Feb-Mar). Taspine isolated from Radix et Rhizoma Leonticis inhibits proliferation and migration of endothelial cells as well as chicken chorioallantoic membrane neovascularisation. *Vascul Pharmacol, 48*(2–3), 129–137. https://doi.org/10.1016/j.vph.2008.01.008

9 Single-Cell RNA Sequencing

Robert Morris and Feng Cheng
University of South Florida

CONTENTS

9.1 Introduction to Single-Cell RNA Sequencing 233
9.2 Microdroplet Approaches to Cell Capture 234
    9.2.1 Drop-Seq Platform 235
    9.2.2 Chromium System 237
9.3 Non-Microfluidic Approaches to Cell Capture 238
    9.3.1 Fluorescence-Activated Single-Cell Sorting (FACS) 238
    9.3.2 CytoSeq 239
    9.3.3 SPLiT-seq 239
9.4 scRNA-seq Output Data 239
    9.4.1 Amplification Step in scRNA-seq 239
    9.4.2 Output Data of scRNA-seq 240
9.5 scRNA-seq Data Analysis 240
    9.5.1 ScRNA-seq Analysis Program 1: Cell Ranger 241
    9.5.2 ScRNA-seq Analysis Program 2: STARsolo 242
    9.5.3 ScRNA-seq Analysis Program 3: DropletUtils 247
    9.5.4 ScRNA-seq Analysis Program 4: Seurat 251
9.6 Limitations of scRNA-seq 257
9.7 Applications of scRNA-seq in Drug Discovery 258
9.8 Summary 259
Keywords and Phrases 259
Bibliography 260

9.1 INTRODUCTION TO SINGLE-CELL RNA SEQUENCING

In previous chapters, we introduced what is collectively referred to as bulk RNA sequencing. In principle, bulk sequencing involves isolation of RNA species, either purely mRNAs or the entire transcriptome of many different kinds of RNAs, from a tissue sample or population of cells. We then typically look for differential expression patterns of genes between samples in response to various treatment paradigms. Overall, bulk sequencing captures the average global expression pattern of transcripts across all cells in the sample. However, for heterogeneous populations of cells, which may arise in tumors or be due to the

DOI: 10.1201/9781003174028-9

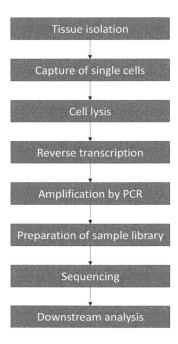

FIGURE 9.1 Overview of the scRNA protocol.

inherent minute differences of cells in a given region due to micro and macro stimuli, bulk RNA sequencing may fail to detect differences in expression attributed to rare populations of cells.

Essentially, single-cell RNA sequencing (scRNA-seq) improves the sensitivity of the experiment and can be used to identify differentially expressed genes in a heterogeneous cell population. The overall premise of scRNA-seq is comparable to the bulk RNA sequencing discussed throughout the rest of the textbook (Figure 9.1). We prepare our samples by isolating RNA, sequence the samples to generate our raw reads, align these reads to a reference genome, and then perform various downstream analyses depending on the goals of our particular experiment. What does differ between more traditional bulk sequencing and scRNA-seq is that individual cells must be captured and isolated from one another as well as the primary focus of downstream analysis on identifying cluster patterns of cells with regard to similar differential expression. We will first discuss the methods in which individual cells can be captured.

9.2 MICRODROPLET APPROACHES TO CELL CAPTURE

Of the two most common approaches in which to capture and isolate individual cells, microdroplet-based approaches are most commonly utilized. In this approach, hundreds of thousands of aqueous microdroplets, each of which is

Single-Cell RNA Sequencing

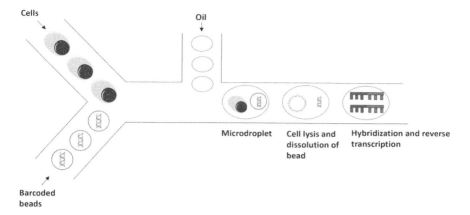

FIGURE 9.2 Overview of the Chromium system microdroplet approach. Individual cells and barcoded beads are introduced into a microfluidics system along with oil. The oil creates small microdroplets in which a cell and barcoded bead are encapsulated into. Next, the cell undergoes lysis and releases mRNA transcripts into the microdroplet while the bead dissolves, the latter of which releases uniquely barcoded oligonucleotides. Finally, the mRNAs undergo hybridization with the oligonucleotide barcode and reverse transcription occurs to generate cDNA.

surrounded by a thin layer of oil, containing a single cell and a hydrogel bead carrying a uniquely barcoded oligonucleotide sequence are first generated (Figure 9.2). Next, the hydrogel bead encapsulated in the microdroplet dissolves and releases the barcoded oligonucleotide sequence. Simultaneously, the cell undergoes lysis and releases mRNA transcripts into the microdroplet, allowing for the incorporation of the unique barcode label into the cDNA created during reverse transcription. In particular, the barcode is incorporated into the poly-A tail of mRNA transcripts. The barcode allows for each transcript to be assigned to a single cell, thus enabling the characterization of cell-specific transcriptomic profiles. Each transcript derived from a single cell will have the same cell barcode (CB).

9.2.1 Drop-Seq Platform

There are two primary platform technologies that utilize this microdroplet approach: The Chromium system and Drop-Seq platform. Although similar, the location in which the hybridization of the barcode oligonucleotide occurs differs. In the Drop-Seq platform, the beads, the barcode oligonucleotides, and the mRNAs in which the barcodes have annealed to are all released into a single tube. Thus, the reverse transcription step does not occur within the microdroplet but in a small container. In contrast, the reverse transcription step occurs directly within each microdroplet as will be discussed in the next section on the Chromium platform.

Drop-Seq was initially developed in 2015 as a means to perform parallel analysis of the transcriptomic profiles of thousands of cells simultaneously through an encapsulation methodology. The standardized protocol for Drop-Seq analysis involves the following steps:

1. Preparation of a single-cell suspension.
2. Encapsulation of a single cell with a unique identifier barcode bead within a nanoliter-scale droplet.
3. Lysis of encapsulated cells.
4. Capturing of mRNA and the formation of single-cell transcriptomes attached to microparticles (STAMPs).
5. Performing a single coupled reaction of reverse transcription, amplification, and sequencing of all STAMPs.
6. Mapping of STAMPs to each individual cell using the barcode.

First, oligonucleotide primers are synthesized directly on the microparticle beads with 5′ to 3′ directionality. Each constructed oligonucleotide contains a constant sequence that is located on each bead that is used for downstream polymerase chain reaction (PCR) and a unique CB that is identical across each primer on a bead yet is different to any other barcode that is used to identify the cell of origin. In addition, each oligonucleotide contains a unique molecular identifier (UMI) sequence that allows for the identification of PCR duplicates as well as an oligo-dT sequence that enables the capturing of mature mRNA sequences for the purposes of reverse transcription. A split-and-pool method is used in order to generate a large supply of beads that each contains a unique barcoding sequence. In this method, a large pool consisting of millions of individual beads is subdivided into four equal-sized groups, which is subsequently followed by the introduction of a different nitrogenous base (A, G, C, or T) to each of the subgroups. After incorporation, the four subgroups are recombined into a single large pool and then randomly re-sorted into four new groups that are each exposed to a new nitrogenous base. After 12 cycles of this synthesis process, the resulting product is a collection of millions of beads that each contains 1 of 4^{12} possible 12-sequence bar codes. Next, the UMIs are generated by eight cycles of degenerate oligonucleotide synthesis. Finally, the oligo-dT sequence required for mature mRNA capture is synthesized on the free 3′ end of the oligonucleotide.

Encapsulation of the cells into these beads is carried out via a microfluidics system that co-flows two aqueous solutions, one of which contains our beads in a lysis buffer and the other of which contains our single-cell suspension, through an oil layer to generate nanodroplets. However, only some of the nanodroplets will be of interest as we are only concerned with the ones that contain both a cell and a bead. Some nanodroplets that are generated using this method may lack both the bead and the single cell or may only have one of the two essential components. In addition, we need to exclude any droplets that contain multiple cells or multiple beads. The proportion of nanodroplets containing one cell and one bead are the STAMPs that will undergo further analysis, be broken down, and

Single-Cell RNA Sequencing

the released captured mature mRNA transcripts captured will undergo reverse transcription.

9.2.2 Chromium System

Like the Drop-Seq methodology, the Chromium 10X system from the Illumina sequencing platform uses a microfluidics system to incorporate a bead containing necessary identification sequences and a single cell from a high-quality suspension into a nanodroplet. As previously discussed, the primary difference is that the reverse transcription step occurs directly within the nanodroplet under the Chromium 10X system, while conversely reverse transcription for the Drop-Seq system occurs after the splitting of the nanodroplet and thus does not occur within the lumen of the droplet. Construction of our barcoded cDNA library from our cells of interest begins with the generation of Gel Beads in Emulsion (GEMs), the reaction vesicles in which the reverse transcription of captured mRNA will take place. 10X Single Cell 3′ Gel Beads, a Master Mix solution of our cells of interest, the Partitioning Oil, and other reagents are incorporated onto the Chromium Chip B. Because the Master Mix of Cells is rather dilute, only a small fraction of GEMs will contain a single cell, while many others will lack any cells. Once the GEMs are assembled, the Gel Bead is dissolved, resulting in the subsequent lysing of the cell contained within the GEM and the release of primers. Each primer contains an Illumina TruSeq Read 1 sequence, a 16-nt identification 10X barcode, a 12-nt UMI sequence, and a 30-nt polydT sequence that is designed to bind and capture mature mRNAs. These primers mixed with a combination of the cell lysate, additional Master Mix, and reverse transcription reagents will generate our library of full-length cDNA containing the barcode sequences following incubation.

Once the GEMs have been incubated, silane magnetic beads are used to purify and isolate the cDNA from leftover reagents and are then amplified using PCR to extend the size of the cDNA library. As per usual with RNA sequencing, the cDNA is subsequently fragmented and then incubated with a combination of the P5 and P7 primers, which will ligate to opposing ends of each fragment and allow for bridge amplification, along with a sample index and the Illumina TruSeq Read 2 sequence. When sequencing, Read 1 will encode the UMI and barcode sequences, while Read 2 will relay the sequence of the bound fragment.

The Chromium system technology has several key advantages over the Drop-Seq platform. First, the Chromium system has reduced technical noise and greater sensitivity due to its ability to detect a higher number of genes per individual cell than the Drop-Seq system, which is especially important when attempting to separate closely related cell types that have subtle differences in their gene expression profiles. Next, the Chromium system is significantly more efficient as it allows for a much higher number of input cells to be incorporated into a microdrop that contains a barcode bead. In contrast, the Drop-Seq system incorporates both the barcoded beads and the individual cells into a microdrop in a random fashion, resulting in many microdrops containing only the cell and lacking the bead encapsulating the barcoded oligonucleotides. The random incorporation of

TABLE 9.1
Comparison of the Drop-Seq and Chromium Microdroplet Systems

Chromium System	Drop-Seq System
Greater sensitivity (more genes per cell detected)	Lower sensitivity (less genes per cell detected)
More deliberate incorporation of beads and cells into microdroplet	Randomized incorporation of beads and cells into microdroplet
>50% success rate of generating RNA-seq usable microdroplets	5% success rate of generating RNA-seq usable microdroplets
More costly due to reagents required	Less costly
Reverse transcription occurs within the microdroplet	Reverse transcription occurs outside the microdroplet in a separate tube

constituents can give rise to two distinct problematic scenarios beyond the generation of microdrops lacking a barcoded bead. In the first scenario, a microdrop may be incorporated with two cells and one barcoded bead, resulting in the transcripts derived from both cells being labeled with the same singular barcode and the merging of the transcripts of two cells into one. In the second scenario, microdrops containing two barcoded beads and one cell would result in transcripts derived from a single cell being labeled as if they were derived from two unique cells (transcripts from the same cell being labeled with one of two possible barcodes). As a result, the protocol for the Drop-Seq system aims to incorporate 1 cell in every 20 microdrops, which results in a frequency of approximately 1/400 with regard to the incorporation of two cells into a single microdrop, as well as one bead per every 10 microdrops. As a result, only approximately 5% of the generated microdrops contain the appropriate number of cells and beads and are usable for RNA sequencing. In contrast, the Chromium system has a success rate of over 50%, and thus, a significant number of usable microdrops are generated.

The primary disadvantage of the Chromium system when compared to the Drop-Seq system is that the reagents required for the Chromium system are more costly than their respective counterparts required by the Drop-Seq system. Comparison of the two systems is presented in Table 9.1.

9.3 NON-MICROFLUIDIC APPROACHES TO CELL CAPTURE

9.3.1 FLUORESCENCE-ACTIVATED SINGLE-CELL SORTING (FACS)

One of the more prominent non-high-throughput approaches to cell capture and isolation is a flow cytometry technique known as fluorescence-activated single-cell sorting (FACS). In this method, antibodies conjugated with a fluorescent molecule (termed fluorochrome) are utilized that bind to intracellular and extracellular proteins of interest with high specificity. The labeled cells that are suspended in a liquid stream then are passed through a laser beam one-at-a-time and

Single-Cell RNA Sequencing

the light scattering pattern as well as the intensity of the fluorescent signal are detected by an electronic light detector. Theoretically, the intensity of fluorescence will correlate with the amount of the cellular flow cell component. The cell sorting machine then imparts an electrical charge on each cell so that each cell can be sorted by charge once they exit the flow chamber.

9.3.2 CYTOSEQ

CytoSeq is another non-microfluidic approach to cell capture that has significantly higher throughput than the aforementioned FACS protocol. In this technology platform, gravity is used to allow individual cells to settle into microwells. Once the single cells are settled, a bead suspension is then introduced at saturation concentration in order to ensure that a bead is distributed to each microwell. Like the beads used in the previously discussed Chromium and Drop-Seq platforms, each bead, each of which is sized so that a max of one bead can be incorporated into each microwell, contains a unique set of barcoded oligonucleotides required for the priming of reverse transcription. Next, excess beads are removed and the cells undergo lysis, releasing mRNA that can subsequently hybridize to the bead-derived oligonucleotide labels.

9.3.3 SPLiT-SEQ

An additional scRNA-seq technology system that yields comparable results to Chromium and Drop-Seq is SPLiT-seq, a cost-effective platform that costs approximately $0.01 for each sequenced cell. In this method, the cells are first gently fixed and permeabilized, at which point they themselves serve as microdrops. This differs from the Chromium and Drop-Seq methods in that oil is not used to construct a microdrop encapsulating both the cell and the barcoded bead. Each cell is then randomly distributed across each well of a 96-well plate and reverse transcription takes place within the cells using primers that are specific for each well. Barcode sequences are then ligated onto the cDNAs in a sequential manner with each cell having a unique barcode oligonucleotide sequence ligated to their respective cDNA molecules. The cells are then pooled together, mixed, and then randomly distributed once more into the wells of a new 96-well plate. Short well-specific oligonucleotides are then introduced and ligated to the cDNA sequences. The process of pooling and splitting is repeated three additional times in order to generate a total of 21,233,664 possible unique barcode sequences.

9.4 scRNA-seq OUTPUT DATA

9.4.1 AMPLIFICATION STEP IN SCRNA-SEQ

The Illumina sequencing platform, among others, commonly utilizes a sequencing-by-synthesis method for sequencing reads derived from single-cell data. Prior to sequencing by the instrument, an amplification step, usually bridge

240 RNA-seq in Drug Discovery and Development

amplification, occurs to generate groups of cDNA fragment clones in each microwell. Because the general mRNA content of a given cell is rather low, we must first amplify our sequences so that the signal is strong enough to be detected by the sequencing instrument.

Adapters and polymerases required for sequencing are incorporated and bound to the bottom of each microwell and subsequently interact with the adapter sequences ligated to the cDNA sequences. The insert sequences then undergo a series of replicate steps that utilizes the polymerases and fluorophore-bound nucleotides to amplify the sequences. During each PCR cycle, a single nucleotide is added, and if it is incorporated by the polymerase, a light emission is released that is detected by the instrument sensor. Each of the four nucleotides elicits a unique light emission that can be differentiated by the sequencing instrument. Once all of the fragments are amplified and the current cycle is completed, the sensor records the light spectra released by the entire fragment library. The sequencer can then reconstruct at the end of each cycle the entire sequence of all of the inserts as well as the barcode tabs in order to assign each sequence to its appropriate library.

9.4.2 OUTPUT DATA OF scRNA-SEQ

Unlike bulk sequencing, sequencing of single-cell data generates raw reads in the form of a binary base call (BCL) file along with their corresponding quality scores. In the case of pooled libraries in which multiple different barcodes are used, the larger BCL file must be demultiplexed so that each read is properly assigned to their sample or cell of origin. These individual BCL files must then be converted into FASTQ files prior to downstream analysis.

Each read sequence contains three crucial components that are required to map and quantify our reads. This includes

1. The cDNA fragment sequence that identifies the particular RNA transcript and what it encodes for.
2. The CB specifies the cell of origin in which the given transcript is expressed.
3. The UMI that identifies the specific RNA molecule (each RNA molecule will have its own UMI).

9.5 scRNA-seq DATA ANALYSIS

The mapping and quantification steps of scRNA-seq can be briefly summarized in four distinct steps:

1. Mapping cDNA fragments to a reference genome.
2. Assigning each read to a particular gene.
3. Assigning these reads to their cells of origin (cell barcode demultiplexing).
4. Counting the number of unique RNA molecules (UMI deduplication).

Single-Cell RNA Sequencing

The primary output generated from the mapping and quantification steps presented above is a gene/cell expression matrix, which contains counts for each gene in each cell. Like in bulk RNA sequencing, a raw expression matrix is a simple table that displays the number of reads that map to a given gene for all cells. In the processed and filtered matrix, the matrix is an I×J table with each row corresponding to a unique gene and columns denoting each cell based on the unique cell barcode.

9.5.1 scRNA-seq Analysis Program 1: Cell Ranger

Cell Ranger is a proprietary set of pipelines from 10X Genomics that is used to process single-cell data derived from the Chromium system of cell capturing. As presented in Figure 9.3, five pipelines are included in Cell Ranger that take care of

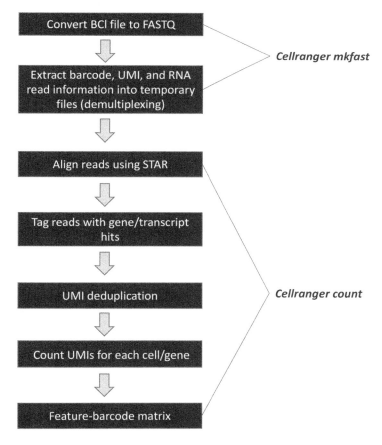

FIGURE 9.3 Overview of the pipelines used by Cell Ranger to generate a feature-barcode matrix.

242 RNA-seq in Drug Discovery and Development

alignment, the formation of a gene/cell expression matrix (sometimes referred to as a feature-barcode matrix), and subsequent clustering and other forms of analysis.

Assuming that we are starting with our raw BCL files produced directly from the sequencing instrument, the five pipelines available in Cell Ranger include

1. **Cellranger mkfast**: This pipeline demultiplexes raw BCL files into FASTQ files. Demultiplexing is defined as the process of sorting a set of sequenced reads into separate files with each file representing a separate sample run.
2. **Cellranger count**: This pipeline takes demultiplexed FASTQ files as input and performs alignment, filtering procedures, counting of the barcodes, and UMI deduplication. This pipeline generates feature-barcode matrices based on the Chromium cell barcodes as well as performs gene expression analysis and cluster analysis. Cluster analysis allows for the grouping of cells that have highly similar expression patterns with individual groups representing a distinct cell phenotype.
3. **Cellranger aggr**: Multiple outputs generated from the *cellranger count* pipeline can be combined into a single file. Simultaneously, each run is normalized by sequencing depth and recomputes feature-barcode matrices from the aggregated data.
4. **Cellranger reanalyze**: This pipeline takes as input feature-barcode matrices generated from either *Cellranger count or Cellranger aggr* and reperforms the dimensionality reduction, clustering analysis, and expression analysis with flexible user-input controls.
5. **Cellranger multi**: For cell multiplexing data, this pipeline takes FASTQ files generated by *Cellranger mkfast* and performs the same steps as *Cellranger count*. Cell multiplexing refers to the mixing of cells that each is labeled with a different molecular barcode.

9.5.2 ScRNA-seq Analysis Program 2: STARsolo

Cell Ranger is currently the most popular commercial single-cell data analysis tool and is widely used in many research fields. However, due to the proprietary nature of the tool suite and its applicability to only single-cell data derived from the Chromium capture system, Cell Ranger is not currently available through the Galaxy interface. Instead, we will introduce another tool that is available through the Galaxy toolshed that can replicate the feature-barcode matrices with high accuracy and sensitivity. STARsolo expands on the base RNA-STAR tool and runs in a manner similar to the use of STAR for bulk RNA sequencing. However, additional features specific to single-cell sequencing including barcode size (whether or not the barcode is the same length as the read), number of expected cells, type of UMI filtering, and deduplication method are also adjustable. In addition, a whitelist of all possible barcodes in the prepared library must be provided as input.

Single-Cell RNA Sequencing

In general, the procedure utilized by STARsolo is similar to the one made used by the Cell Ranger algorithms. This includes

1. Aligns reads in FASTQ format to the reference genome and assigns each read to a gene based on concordance with transcript models. Concordance of alignment is considered to be sufficient if all of the alignment blocks lie within the exons of the transcript and are in agreement with splice junctions if they are present.
2. Cell barcodes are matched and assessed for concordance with the provided whitelist using a simple binary search algorithm.
3. CB, UMI, and gene sequences/counts are stored in separate temporary files.
4. CBs are demultiplexed (separation of our read sequences to their respective sample) and error correction occurs.
5. UMIs are deduplicated and counted for each gene/genomic feature.
6. Raw feature-barcode (gene-cell count matrix) is generated.
7. Filtering removes ambient RNA molecules. Ambient mRNA refers to extraneous RNA molecules released into suspension by stressed or apoptotic cells. Ambient mRNA serves as cross-contamination and possible experimental noise.
8. Generation of the filtered feature-barcode matrix.

We will use a public dataset provided by 10X Genomics in order to run STARsolo in the Galaxy interface. The dataset is derived from 1,000 peripheral blood mononuclear cells and can be accessed and directly downloaded from the Zenodo repository using the following link: https://zenodo.org/record/3457880#. YoG8dOjMJPY. However, due to the extensive memory required to run STARsolo, we will map the cDNA reads to only the x chromosome. To successfully run STARsolo in Galaxy, we will need to download the following files:

- subset_pbmc_1k_v3_S1_*L001*_R1_001.fastq.gz
- subset_pbmc_1k_v3_S1_*L001*_R2_001.fastq.gz
- subset_1k_v3_S1_*L002*_R1_001.fastq.gz
- subset_1k_v3_S1_*L002*_R2_001.fastq.gz
- Homo_sapiens.GRCh37.75.gtf
- 3M-february-2018.txt.gz

The first four files correspond to our read files that are split into lanes 1 (*L001*) and 2 (*L002*). Reads labeled *R1* contain reads of our barcode sequences, while the files labeled *R2* denote reads for the cDNA sequences. *The Homo_sapiens.GRCh37.75. gtf* file is the reference genome file in GTF format and the *3M-february-2018.txt. gz* contains the whitelist of all barcode sequences.

Once the six files have been imported into our Galaxy history, we need to determine the underlying library chemistry that was used to construct our reads (essentially we need to determine which primers were used and the number of

FIGURE 9.4 Parameters utilized to generate STARsolo output files.

base pairs present for the CB and UMI combined). For v2 chemistry, the length of the CB is 16 bp and the length of the UMI is equal to 10 bp (total = 26 base pairs). In contrast, libraries with v3 chemistry have a total length of 28 bp (CB = 16 bp; UMI = 12 bp). This information is necessary for the deduplication and demultiplexing steps of single-cell data processing. In this instance, our samples were constructed using v3 chemistry. We can determine this by simply viewing one of the *R1* files in Galaxy and counting the number of base pairs in a single read.

Single-Cell RNA Sequencing   245

Figure 9.4 illustrates the parameters utilized to generate our sample output and includes the following:

1. Set *custom or built-in reference genome* option to use a built-in index.
2. Set *reference genome with or without an annotation* option to use genome reference without built-in gene model.
3. Set *select reference genome* option to Human (Homo Sapiens): hg 19 chrX.
4. Set *gene model (gff3, gtf) file for splice junctions* option to our GTF file Homo_sapiens.GRCh37.75.gtf.
5. Set *length of genomic sequence around annotated junctions* option to 100.
6. Set *type of single-cell RNA-seq* option to Drop-Seq or 10X Chromium.
7. For the *RNA-seq FASTQ/FASTA file, Barcode reads* option, select our two R1 read files using the Ctrl key.
8. For the *RNA-seq FASTQ/FASTA file, cDNA reads* option, select our two R2 read files using the Ctrl key.
9. Set *RNA-Seq cell barcode whitelist* option to our text file 3M-february-2018.txt.gz.
10. Set *configure chemistry options* to Cell Ranger v3.
11. Set *UMI deduplication (collapsing) algorithm* to collapse all UMIs with 1 mismatch distance to each other.
12. Set *Matching the cell barcodes to the whitelist* option to multiple matches (CellRanger 2).
13. Set *cell filter type and parameters* to do not filter.
14. Leave the remaining options to their respective default settings.

STARsolo outputs six unique files including a program log, a summary of mapping quality, alignments in BAM format, and three count matrix files (Matrix Gene Counts, Barcodes, Genes). Figure 9.5 shows a sample excerpt from the raw matrix gene counts file output generated after completion of STARsolo. In addition, Figure 9.6 displays summary statistics for the barcodes and gene features.

At this stage, there are three statistics parameters that we should take notice of from the summary report: *nCellBarcodes, nMatch,* and *nNoFeature.* The *nCellBarcodes* statistic tells how many cells/unique cell barcodes were detected (80,373), the *nMatch* statistic denotes the number of proper alignments (5,181,294), and the statistic *nNoFeature* tells us how many reads mapped to the genome but did not have an associated annotation. Ideally, we want the *nMatch* statistic to be quite large as it shows that most of the reads were aligned to the reference genome and we want the *nNoFeature* statistic to be relatively small. However, as we aligned reads comprising all of the genome while using an annotation file just for chromosome X, it is expected that the value of *nNoFeature* would be larger than usual. If we look at the raw matrix generated by STARsole (no filtering), we can observe many instances of low counts (lots of counts are equal to 0 or 1). This will be of importance for the next section in which we use a new tool in order to eliminate low-quality cells and produce a higher quality matrix.

```
%%MatrixMarket matrix coordinate integer general
%
        2392           6794880            327300
        1811                24                 1
        1903               567                 1
        1811               570                 1
         138               677                 1
        1274               830                 1
          11              1030                 4
          13              1030                 3
          22              1030                 2
          26              1030                 1
          41              1030                 1
          53              1030                 1
          57              1030                24
          62              1030                 8
          76              1030                 5
          79              1030                 1
          80              1030                 1
         101              1030                 2
         117              1030                 1
         124              1030                 2
         130              1030                 3
         133              1030                 1
         137              1030                 2
         138              1030                75
         147              1030                 2
         149              1030                20
         152              1030                 1
         165              1030                 2
         167              1030                10
         182              1030                10
         188              1030                14
         221              1030                 1
         233              1030                 1
         239              1030                 1
         243              1030                 1
         247              1030                 2
         255              1030                 2
         259              1030                 1
         278              1030                 1
         283              1030                12
         285              1030                18
         287              1030                 1
```

FIGURE 9.5 Excerpt of the gene-count matrix generated as output by STARsolo.

Single-Cell RNA Sequencing

```
Barcodes:
                           nNoAdapter              0
                              nNoUMI               0
                               nNoCB               0
                              nNinCB               0
                             nNinUMI            3484
                      nUMIhomopolymer           45569
                            nTooMany               0
                             nNoMatch         1476419
               nMismatchesInMultCB              0
                          nExactMatch        63205957
                     nMismatchOneWL           402565
                    nMismatchToMultWL         1467893
Genes:
                            nUnmapped        49201263
                           nNoFeature        10565546
                        nAmbigFeature          116976
                nAmbigFeatureMultimap           62536
                             nTooMany            7272
                       nNoExactMatch             4064
                          nExactMatch         5069427
                               nMatch         5181294
                        nCellBarcodes           80373
                                nUMIs         1699586
```

FIGURE 9.6 Summary statistics for the matrix gene counts output from STARsolo.

9.5.3 ScRNA-seq Analysis Program 3: DropletUtils

The three matrices generated by STARsolo in the above section are formatted as bundles. This means that the information we need to construct a tabulated matrix for Genes vs. Cells are separated into different files. The above raw matrix contains many gene features and thousands of cells. If we attempt to construct the new tabulated matrix directly, we will generate a very large Genes vs. Cells matrix that is very sparse. In addition, the raw matrix contains many low-quality cells as well as wells containing just a barcode and not an actual cell (empty droplets). This phenomenon explains the large discrepancy in the number of starting cells ($n = 1,000$) and the number of reported cell barcodes in the STARsolo summary statistics. In addition, a process known as barcode switching may also generate erroneous results. This occurs because a washing step does not occur between steps in the flow cell and because flow cell seeding as well as DNA amplification occurs simultaneously. As a result, it is possible for extraneous free-floating barcodes that are still in the solution to be extended using DNA from different libraries. This would give rise to a mixing of transcriptomes and an assignment of transcripts to cells in which they did not originate from. It is estimated that this occurs in approximately 2.5% of total reads.

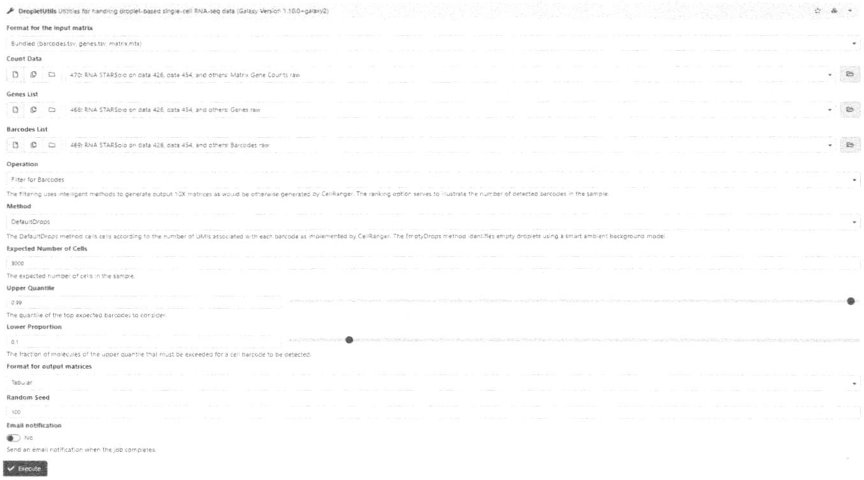

FIGURE 9.7   Parameters used for the *filter for barcodes* operation in DropletUtils.

The DropletUtils tool is used to remove erroneous cells or false positives as well as to eliminate low-quality cells. The following steps, displayed in Figure 9.7, were used to generate the higher quality Genes vs. Cells matrix that closely resembles the output that would be generated from the Cell Ranger pipeline:

1. Set *format for the input matrix* to bundled (barcodes.tsv, genes.tsv, matrix.mtx).
2. For *count data*, select matrix gene counts raw produced from STARsolo as input for this option.
3. For *genes list*, select genes raw produced from STARsolo as input for this option.
4. For *barcodes list,* select barcodes raw produced from STARsolo as input for this option.
5. Set *operation* to filter for barcodes.
6. Set *method* to DefaultDrops. This method calls cells based on the number of UMIs associated with each cell barcode.
7. Set *format for output matrices* to tabular.
8. Leave the remaining options to their respective default settings.

Figure 9.8 presents a sample excerpt from the higher quality matrix produced by the DropletUtils tool. Each row represents a gene feature and each column corresponds to a unique cell barcode. Each count for a given position corresponds to the number of transcripts matching the annotation for each gene in each individual cell.

Another method that the DropletUtils tool can use to assess and identify high-quality cells is through the *Rank Barcodes* option, which requires the same inputs

Single-Cell RNA Sequencing

FIGURE 9.8 Sample output of the higher quality Genes vs. Cells matrix produced by the DropletUtils tool. Genes are denoted by each row and each column denotes each cell by their respective unique barcode.

FIGURE 9.9 Parameters used for the *rank barcodes* operation in DropletUtils.

as described above for the *DefaultDrops* tool (Figure 9.9). The purpose of this tool is to estimate the expected minimum number of transcript barcodes (UMIs) per cell by generating a gap statistic plot (Figure 9.10). In the gap statistic plot, the y-axis denotes the log total number of cellular barcodes that each corresponds to an individual cell, while the x-axis denotes the minimum number of transcripts expected in each cell. The 'knee' line denotes the number of expected cells we should have based on our data input, while the inflection point represents the transition from low-quality cells to high-quality cells as defined by the value we

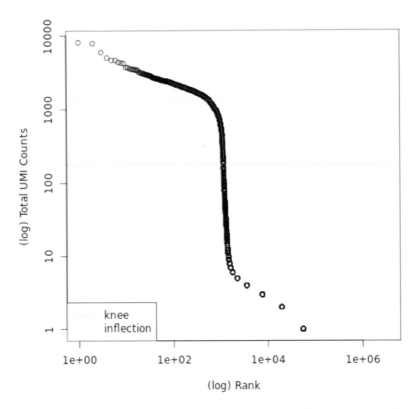

FIGURE 9.10 The gap statistic graph generated by the *rank barcodes* operation of DropletUtils. Important features include the knee and inflection points (transition between high- and low-quality cells) as well as the dropoff point (minimum expected number of unique UMIs per high-quality cell).

assign to the *lower bound* parameter when running the *Rank Barcodes* option (default = 100 minimum UMIs).

In our example, the knee line is equal to 1,321, which is reasonably in accordance with our initial starting point of 1,000 cells, and the inflection point occurs at 180. Anything above the inflection point is assumed to represent empty cells and/or cells with very low RNA content (UMI below the threshold). The dropoff point between the two transitions estimates the minimum number of UMIs that are expected to be detected in each high-quality cell. In this instance, the dropoff indicates approximately 300 UMIs as the minimum number of transcripts present in each high-quality cell as indicated by the value that we assign to the *upper quintile* parameter when running the *filter for barcodes* operation. In the next sections, we will discuss tools that can be used to enhance visualization and manipulation of single-cell data.

Single-Cell RNA Sequencing

9.5.4 ScRNA-seq Analysis Program 4: Seurat

Seurat is a toolkit that allows for greater exploration of single-cell data by generating multiple figures that can identify markers associated with heterogeneous cell populations during differential expression. The algorithms comprising the Seurat toolkit can be summarized in the following five steps:

1. Uses canonical correlation analysis to construct a shared gene correlation structure. This assesses the correlation between two variables, namely, cells and genes in this instance.
2. Identifies individual cells that are poorly described by the shared gene correlation structure to highlight possible rare cell populations.
3. Uses dimensionality reduction techniques to allow for clustering analysis.
4. Performs clustering analysis to group cells that have similar expression patterns.
5. Performs comparative analysis of gene expression within each cluster of cells.

Using the default parameters of Seurat in Galaxy, which are presented in Figure 9.11, as well as the high-quality matrix we produced in the previous

FIGURE 9.11 Parameters used to generate plots for our high-quality count matrix outputted from DropletUtils using the Seurat toolkit.

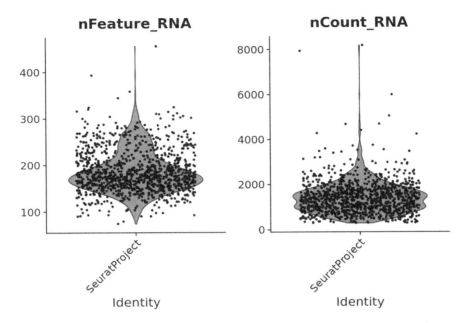

FIGURE 9.12 Sample output from Seurat toolkit: violin plots. In these violin plots, the y-axis plots the log-normalized expression values and the identity classes (i.e. different cell populations) are plotted on the x-axis. By default, Seurat utilizes the term 'identity' when different classes are not specified.

section using the DropletUtils tool, we will generate multiple figures including violin plots, scatter plots, feature-counts plots, principal component (PC) plots, t-Distributed Stochastic Neighbor Embedding (t-SNE) plots, and heatmaps (Figures 9.12–9.20).

Violin plots (Figure 9.12) are a hybrid of traditional box plots and a density plot that shows the probability density of our data. In this case, Seurat generates two separate violin plots; one for the mapped features and one for the counts. For both graphs, most counts and features are clustered toward the bottom of the graph, which is in agreement with our previous analysis. The cell type that we utilized in the experimental design naturally has low RNA content and we only aligned to the X chromosome for the sake of computational efficiency.

Seurat next generates a feature-counts figure (Figure 9.13) that plots the *nFeature_RNA* on the y-axis and the *nCount_RNA* on the x-axis as well as calculates the shared correlation (canonical correlation analysis=0.83 for this example). In addition, Seurat generates a figure that plots the dispersion estimates against the average expression of each detected gene in order to find statistically significant differentially expressed genes (Figure 9.14). In this figure, red dots represent genes that achieved statistical significance and are differentially expressed.

Single-Cell RNA Sequencing

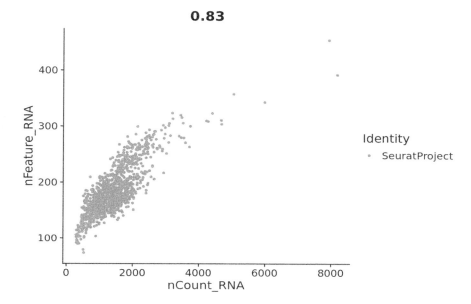

FIGURE 9.13 Sample output from Seurat toolkit: Feature-counts plots. These graphs plot the feature counts on the y-axis and the raw counts on the x-axis as well as calculate the strength of the correlation between the two measures.

Seurat then performs dimensional reduction analysis and displays this information in various PC plots by reducing the number of variables (Figures 9.15–9.17). In principal component analysis, we are trying to reduce the complexity of the data and improve our ability to interpret the data by finding several factors that explain most of the variation in values we observed. This statistical method allows us to cluster our many datapoints into a few significant groups, thus improving our ability to visualize the similarities in datapoints, i.e., expression of each gene. By default, Seurat determines the top 10 PCs and generates a graph displaying the standard deviations for each PC (Figure 9.18).

As you can see from Figure 9.17, only five of the ten PCs (PC1, PC2, PC3, PC8, and PC9) achieve statistical significance at an alpha level of 0.05. Thus, these five PCs are the ones utilized to perform the clustering analysis (Figure 9.19) and the differential expression analysis (Figure 9.20). The tSNE plot presented in Figure 9.19 provides an estimate of the spatial arrangement of each cell based upon the similarities in transcriptome profiles (clustering based on similar PC values). Finally, the expression patterns of each of the statistically significant genes are displayed in the heatmap presented in Figure 9.20.

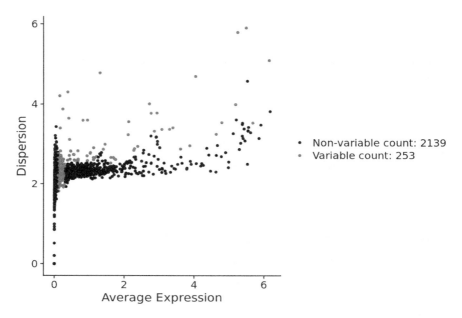

FIGURE 9.14 Sample output from Seurat toolkit: The dispersion estimates against the average expression of each detected gene.

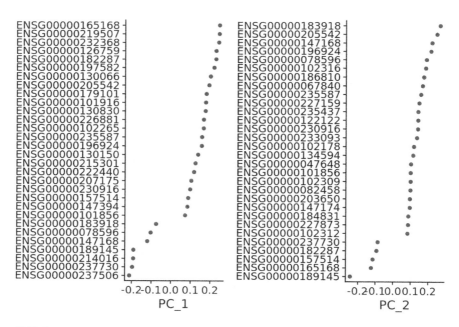

FIGURE 9.15 Sample output from Seurat toolkit: Sample list of genes and the degree in which the given PC explains in.

Single-Cell RNA Sequencing 255

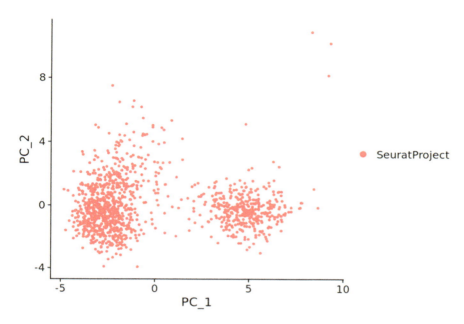

FIGURE 9.16 Sample output from Seurat toolkit: First two principal component analysis components tested by Seurat.

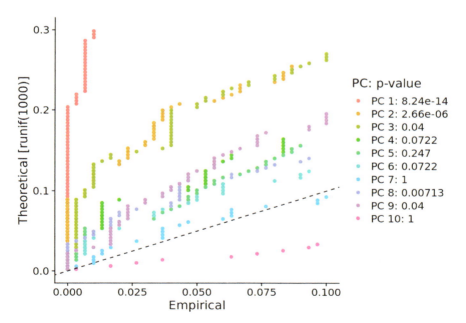

FIGURE 9.17 Sample output from Seurat toolkit: Determination of statistically significant PCs.

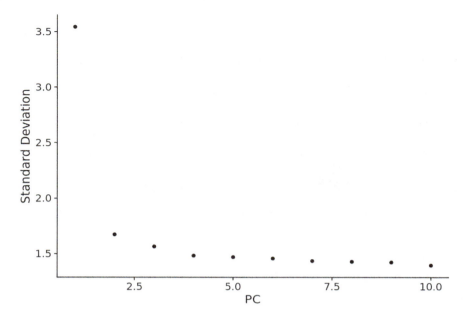

FIGURE 9.18 Sample output from Seurat toolkit: The standard deviations for each PC.

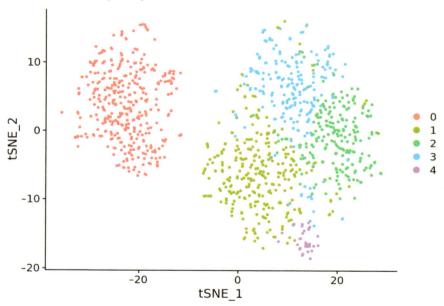

FIGURE 9.19 Sample output from Seurat toolkit: tSNE plot. A tSNE plot is a visualization tool that assesses the similarities between cells in a high-dimensional space and then displays the similarities in a plot with lower dimensionality. This allows for the simplification of data presenting by grouping datapoints based on similarities.

Single-Cell RNA Sequencing

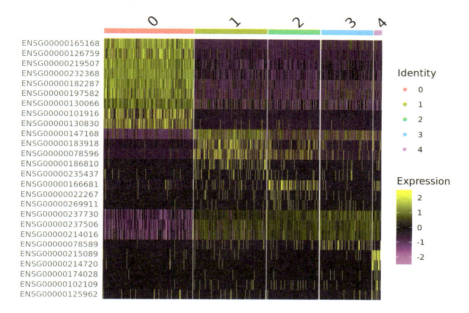

FIGURE 9.20 Sample output from Seurat toolkit: Heatmap of the expression levels of each of the statistically significant genes. Each cell population is denoted by the identity key.

9.6 LIMITATIONS OF scRNA-seq

Despite the promising aspects of scRNA-seq and the in-depth information it can provide with regard to differential gene expression and clustering of cells with similar transcriptomes, scRNA-seq presents several limitations. First, the amount of workable RNA samples is very small as the total RNA in each cell is approximately 10 pg and only a small fraction of that sample is composed of mRNA. The necessity of a cDNA amplification step in order to generate a large enough sample to work with can lead to a disproportionate representation of all of the cDNA in a given cell due to its non-linear nature. This introduced bias can be somewhat reduced through the use of a compression step that counts all cDNA reads with the same UMI as a single hybridization event.

Another limitation is that many methods that exist to isolate individual cells such as micromanipulation and laser capture microdissection are effective yet very time-consuming and low-throughput. In addition, laser capture microdissection is difficult to use without damaging the RNA or collecting cellular debris from neighboring cells. FACS is a high-throughput method to isolate individual cells but is most effective when the cell type of interest is rare. A high-throughput method that utilizes currently available scRNA platforms to examine millions of cells simultaneously does exist; however, it cannot be used for rare cell

TABLE 9.2

Comparison of the Advantages and Disadvantages of using a Short Barcode Sequence Relative to a Long Barcode Sequence

Short Barcodes	Long Barcodes
Smaller number of possible barcode sequences available	Larger number of possible barcode sequences available
More resistant to sequencing errors	More prone to sequencing errors
Can use a smaller editing distance	Requires significantly larger editing distances
Cannot handle the use of larger editing distances	Can handle the use of larger editing distances

populations and you lose all spatial information regarding cell placement in the tissue sample. This thus requires a separate step that reconstructs a 3D image of the tissue by using bioinformatic tools to assign positions to each cell based on their gene expression signatures. Several sources of noise exist that can greatly influence the results of scRNA-seq experiments. Approximately 80% of all noise is thought to be attributable to technical limitations. The remaining 20% of noise can be attributed to biological fluctuations in gene expression in all cells that are always variable as well as the low sensitivity of scRNA-seq to detect the entire population of mRNA. It is predicted that only 10%–20% of present mRNA is actually detected using the existing scRNA-seq technology.

Finally, sequencing errors may result in misassignment of cellular barcodes and loss of genomic information of the transcriptomic profile of a given cell. For every 1 bp sequencing error we are trying to protect against, we must space each barcode sequence by 2 bp. For a given barcode length N and edit distance E, the number of barcodes is denoted by the following equation:

$$\# \text{ of barcodes} = 4^{N-(E-1)} \tag{9.1}$$

Usually, we restrict our barcodes to contain only the four naturally occurring nucleotides A, G, C, and T. As a result, the maximum available number of barcode sequences is limited by the length of the barcode sequence as well as the editing distance between barcodes. Ultimately, there are advantages and disadvantages for the use of short or long barcode sequences, which are summarized in Table 9.2. As a result, it is important to take these limitations under consideration when interpreting experimental data and drawing further conclusions.

9.7 APPLICATIONS OF scRNA-seq IN DRUG DISCOVERY

Multiple applications and usages of scRNA-seq exist for drug discovery and related fields of science. For instance, Aissa *et al.* used scRNA-seq to identify subpopulations of cells that develop drug resistance to tyrosine kinase inhibitors in patients with cancer presenting with somatic mutations in tyrosine kinases.

Single-Cell RNA Sequencing   259

In addition, they sought to identify markers of cells associated with their acquired drug resistance and to find drug combinations that can safely and effectively eliminate these persister cell subpopulations. In this study, Drop-Seq was used to isolate individual cells and Seurat was used to visualize the clustering analysis and PCs. 62 unique markers were identified and mapped to 36 genes previously linked with drug tolerance in non-small cell lung cancer tumors, many of which were mapped to NF-κB and MAPK signaling. Drug-tolerant cells were shown to have dysregulated epigenetic modifications, increased epithelial mesenchymal transition, increased tissue development, and decreased cholesterol metabolism. A strong correlation was determined between an increase in presentation of drug tolerant-associated biomarkers and poor patient outcomes. Overall, this study found that the acquisition of drug tolerance is a stepwise process and different subpopulations of cells with varying degrees of drug tolerance can be present in the same tumor.

A study conducted by Shin *et al.* sought to establish a new method that could multiplex cells treated with different drugs and process them simultaneously to improve efficiency and speed. This method involves the transient transfection of a series of short barcoding oligonucleotides, which consists of a sample barcode sequence and a poly-A tail, that can be used to identify samples based upon their drug treatment status. This method was tested using a drug-screening process of 48 drugs in both K562 cells (erythroleukemia) and A375 cells (melanoma). Individual cells were isolated using the Drop-Seq method and processing and cluster analysis was performed using Seurat. Ultimately, 3,091 high-quality cells expressing a minimum of 500 transcripts (UMI$\geq$500) were detected. This study thus designed an effective alternative to the previous methods described in identifying the cell-by-cell transcriptomic responses to various drug exposures.

9.8   SUMMARY

In this chapter, we introduced the emerging field of scRNA-seq and how it differs from the more traditional bulk RNA sequencing techniques discussed in previous chapters. We reviewed both droplet-based and non-droplet-based isolation methods to capture individual cells. The proprietary pipeline suite created by 10X Genomics was introduced and comparable output data was produced using a combination of STARsolo and DropletUtils in Galaxy. We then introduced Seurat, a visualization tool that can take a high-quality Cell-Gene matrix and produce various figures including feature-counts graphs, PC graphs, violin plots, tSNE clustering plots, and heatmaps. Finally, we discussed two applications of scRNA-seq through an overview of recent literature publications.

KEYWORDS AND PHRASES

After reading this chapter, you should be able to demonstrate familiarity with the following words and phrases:

260 RNA-seq in Drug Discovery and Development

- Understand what scRNA sequencing is and how it differs from bulk sequencing.
- Be able to describe the similarities and differences in droplet-based and non-droplet-based isolation methods.
- Understand the three main components of each read: cell barcode, UMIs, and the sequence itself.
- Be able to explain the four major steps of scRNA sequencing including demultiplexing and deduplication.
- Be able to describe the five pipelines of Cell Ranger.
- Be able to replicate the production of a raw Cell-Gene matrix using STARsolo in Galaxy.
- Understand how to create a high-quality matrix using DropletUtils.
- Understand how to perform clustering analysis and create other visualizations/graphs using Seurat.
- Understand the limitations that currently exist for scRNA sequencing.

BIBLIOGRAPHY

Aissa, A. F., Islam, A., Ariss, M. M., Go, C. C., Rader, A. E., Conrardy, R. D., Gajda, A. M., Rubio-Perez, C., Valyi-Nagy, K., Pasquinelli, M., Feldman, L. E., Green, S. J., Lopez-Bigas, N., Frolov, M. V., & Benevolenskaya, E. V. (2021, Mar 12). Single-cell transcriptional changes associated with drug tolerance and response to combination therapies in cancer. *Nat Commun, 12*(1), 1628. https://doi.org/10.1038/s41467-021-21884-z

Batut, B., Hiltemann, S., Bagnacani, A., Baker, D., Bhardwaj, V., Blank, C., Bretaudeau, A., Brillet-Gueguen, L., Cech, M., Chilton, J., Clements, D., Doppelt-Azeroual, O., Erxleben, A., Freeberg, M. A., Gladman, S., Hoogstrate, Y., Hotz, H. R., Houwaart, T., Jagtap, P., Lariviere, D., Le Corguille, G., Manke, T., Mareuil, F., Ramirez, F., Ryan, D., Sigloch, F. C., Soranzo, N., Wolff, J., Videm, P., Wolfien, M., Wubuli, A., Yusuf, D., Galaxy Training, N., Taylor, J., Backofen, R., Nekrutenko, A., & Gruning, B. (2018, Jun 27). Community-driven data analysis training for biology. *Cell Syst, 6*(6), 752–758. https://doi.org/10.1016/j.cels.2018.05.012

Butler, A., Hoffman, P., Smibert, P., Papalexi, E., & Satija, R. (2018, Jun). Integrating single-cell transcriptomic data across different conditions, technologies, and species. *Nat Biotechnol, 36*(5), 411–420. https://doi.org/10.1038/nbt.4096

Danielski, K. (2023). Guidance on processing the 10x genomics single cell gene expression assay. *Methods Mol Biol, 2584*, 1–28. https://doi.org/10.1007/978-1-0716-2756-3-1

Griffiths, J. A., Richard, A. C., Bach, K., Lun, A. T. L., & Marioni, J. C. (2018, Jul 10). Detection and removal of barcode swapping in single-cell RNA-seq data. *Nat Commun, 9*(1), 2667. https://doi.org/10.1038/s41467-018-05083-x

Janjic, A., Wange, L. E., Bagnoli, J. W., Geuder, J., Nguyen, P., Richter, D., Vieth, B., Vick, B., Jeremias, I., Ziegenhain, C., Hellmann, I., & Enard, W. (2022, Mar 31). Prime-seq, efficient and powerful bulk RNA sequencing. *Genome Biol, 23*(1), 88. https://doi.org/10.1186/s13059-022-02660-8

Kaminow B., Yunusov D., Dobin A. (2021). STARsolo: Accurate, fast and versatile mapping/quantification of single-cell and single-nucleus RNA-seq data. bioRxiv 2005, 442755. https://doi.org/10.1101/2021.05.05.442755

Single-Cell RNA Sequencing

Li, X., & Wang, C. Y. (2021, Nov 15). From bulk, single-cell to spatial RNA sequencing. *Int J Oral Sci, 13*(1), 36. https://doi.org/10.1038/s41368-021-00146-0

Liao, X., Makris, M., & Luo, X. M. (2016, Nov 4). Fluorescence-activated cell sorting for purification of plasmacytoid dendritic cells from the mouse bone marrow. *J Vis Exp, 117*. https://doi.org/10.3791/54641

Macosko, E. Z., Basu, A., Satija, R., Nemesh, J., Shekhar, K., Goldman, M., Tirosh, I., Bialas, A. R., Kamitaki, N., Martersteck, E. M., Trombetta, J. J., Weitz, D. A., Sanes, J. R., Shalek, A. K., Regev, A., & McCarroll, S. A. (2015, May 21). Highly parallel genome-wide expression profiling of individual cells using nanoliter droplets. *Cell, 161*(5), 1202–1214. https://doi.org/10.1016/j.cell.2015.05.002

Potter, S. S. (2018, Aug). Single-cell RNA sequencing for the study of development, physiology and disease. *Nat Rev Nephrol, 14*(8), 479–492. https://doi.org/10.1038/s41581-018-0021-7

Shin, D., Lee, W., Lee, J. H., & Bang, D. (2019, May). Multiplexed single-cell RNA-seq via transient barcoding for simultaneous expression profiling of various drug perturbations. *Sci Adv, 5*(5), 249. https://doi.org/10.1126/sciadv.aav2249

Tekman, M., Batut, B., Ostrovsky, A., Antoniewski, C., Clements, D., Ramirez, F., Etherington, G. J., Hotz, H. R., Scholtalbers, J., Manning, J. R., Bellenger, L., Doyle, M. A., Heydarian, M., Huang, N., Soranzo, N., Moreno, P., Mautner, S., Papatheodorou, I., Nekrutenko, A., Taylor, J., Blankenberg, D., Backofen, R., & Gruning, B. (2020, Oct 20). A single-cell RNA-sequencing training and analysis suite using the Galaxy framework. *Gigascience, 9*(10), 1–9. https://doi.org/10.1093/gigascience/giaa102

Zheng, G. X., Terry, J. M., Belgrader, P., Ryvkin, P., Bent, Z. W., Wilson, R., Ziraldo, S. B., Wheeler, T. D., McDermott, G. P., Zhu, J., Gregory, M. T., Shuga, J., Montesclaros, L., Underwood, J. G., Masquelier, D. A., Nishimura, S. Y., Schnall-Levin, M., Wyatt, P. W., Hindson, C. M., Bharadwaj, R., Wong, A., Ness, K. D., Beppu, L. W., Deeg, H. J., McFarland, C., Loeb, K. R., Valente, W. J., Ericson, N. G., Stevens, E. A., Radich, J. P., Mikkelsen, T. S., Hindson, B. J., & Bielas, J. H. (2017, Jan 16). Massively parallel digital transcriptional profiling of single cells. *Nat Commun, 8*, 14049. https://doi.org/10.1038/ncomms14049

Index

absorption, distribution, metabolism, and excretion (ADME) 173–174
adaptors 6–7
alignment
    de novo 53–57
    genome-guided 36–53
    spliced aligners 43–53
    unspliced aligners 36–43
alternative splicing 104–106
American Standard Coding for Information Interchange (ASCII) 13–14
ARF file 159–161

Ballgown 79
Basic Local Alignment Search Tool (BLAST) 212–214
bigwig file 15–17
binary alignment and map (BAM) file 15–17
binary base call (BCL) file 240–242
Bowtie 37–40
bridge read 142–145
browser extendible data (BED) file 17–19
Burrows-Wheeler transform (BWT) 36–38
    Bowtie 37–43
    introduction 37–38

cDNA
    library preparation 4–6
    reverse transcription 6–7
cell barcode (CB)
    demultiplexing 240–242
    introduction 240–244
    long vs. short 258
    v2 chemistry 244
    v3 chemistry 244
cell ranger 241–242
ChimPipe 137–138
Chromium 10X 237–238
cis-splicing 132
Comparative Toxicogenomics Database (CTD) 188–197
constitutive splicing 101–104
context likelihood of relatedness (CLR) 184–186
contiguous segment (Contig) 10
Cuffdiff 2 106–109
Cytoseq 239

D statistic 45
Database for Annotation, Visualization, and Integrated Discovery (DAVID) 91–95
de Bruijn graph 54–56
demultiplexing 240–244
de-novo assembly
    introduction 53–55
    Trinity 55–57
DESeq2 68–75
DEXSeq 112–118
differential gene expression analysis
    Ballgown 79
    DESeq2 68–75
    EdgeR 75–79
differential splicing analysis
    Cuffdiff 2 106–109
    DEXSeq 112–118
    Diffsplice 109–112
    EdgeR 118–122
differentially expressed genes (DEGs) 2
diffsplice 109–112
DropletUtils 247–250
Drop-Seq 235–237
DrugMatrix 198–199

edgeR 75–79, 118–122
Euclidean distance 86, 183–184
EXACTMATCH 38–40
exon 101–104
exonic splicing enhancers (ESEs) 105–106
exonic splicing silencers (ESSs) 105–106

false discovery rate (FDR) 181–182
FASTA file 2, 21
FASTQ file 12–15
FASTQC 2, 21–29
featurecounts 64–67
fluorescence in situ hybridization (FISH) 136
fluorescence-activated single cell sorting (FACS) 238–239
FM indices 48–49
fragments per kilobase per million (FPKM) 2, 68
fusion gene
    fusion RNA transcripts 131–132, 133–135
    generation of fusion genes 130–131

fusion gene (*cont.*)
    guided detection 135–136
    high-throughput detection 136–149
    introduction 130

Galaxy 20
gapped read 35–36
GC content 22–25
gel beads in emulsion (GEMS) 237
gene transfer format (GTF) file 15–17
genome-guided assembly
    spliced aligners 47–53
    unspliced aligners 36–43
GFusion 139–142

heatmaps 86–90
heterogeneous nuclear ribonucleoproteins
    (hnRNPs) 106
hierarchical graph FM index (HGFM) 48
hierarchical indexing for spliced alignment of
    transcripts 2 (HISAT2) 47–50
high-throughput sequencing 2

Illumina 2500 10–11
InFusion 142–145
initially unmapped reads (IUMs) 43–44
integrative Genomics Viewer (IGV) 81–84
intron 101–104
intronic splicing enhancers (ISEs) 105–106
intronic splicing silencers (ISSs) 105–106
ion-sensitive field-effective transistor
    (ISFET)11
Ion Torrent 318 11

Japanese Toxicogenomics Project (TGP)
    196–198

linear models for microarray data (LIMMA)
    123–126
Lipinski's Rule of 5 (RO5) 210
log fold change (LFC) 70
long non-coding RNAs (lncRNAs) 2, 133–135,
    156–157

mapping quality score (MQS) 43
marker-assisted selection (MAS) 219
maximal mappable prefix (MMP) 50–51
messenger RNA (mRNA) 2
microarray quality control (MAQC)
    consortium 200–201
microdroplet cell capture
    Chromium 10X 237–238
    Drop-Seq 235–237
    introduction 234–235
MicroRNA (miRNA) 153–156
microsatellites 219–220

MiRDeep2 157–165
mutual information (MI) 183

negative binomial distribution 69
non-coding RNAs (ncRNAs) 2, 153–157
non-microdroplet cell capture
    Cytoseq 239
    fluorescence-activated single cell sorting
        (FACS) 238–239
    SPLiT-Seq 239
normalization 3, 67–68

overdispersed poisson distribution 76
Oxford Nanopore Technology (ONT) 12

Pearson's correlation coefficient 184
Phred quality Score (Q_{score}) 12–15
poly-A selection 6
Principal Component Analysis (PCA) 71–74,
    252–253
pseudocounts 77
Pubchem 207–209

quantile-adjusted conditional maximum
    likelihood (qCML) 77

reads
    paired-end reads 8–9
    single-end reads 8
reads per kilobase per million (RPKM)
    3–4
ribosomal RNA (rRNA) 2
RNA
    fragmentation 6
    isolation 4–5
    overview 2–4
    selection and depletion 6
RNA binding proteins (RBPs) 105–106
RNA integrity number (RIN) 4–5
RNA sequencing
    platforms 9–12
Roche 454 9

seed 39–40
seed-based aligners
    subread 40–43
Sequence Alignment Map (SAM) file 15
serine-arginine-rich (SR) proteins 105–106
Seurat 251–257
Single Nucleotide Polymorphisms (SNPs) 2
single-cell RNA sequencing (scRNA-seq)
    applications 258–259
    cell capture 234–239
    data Analysis 239–257
    introduction 233–234
    limitations 257–258

Index

single-cell transcriptomes attached to
microparticles (STAMPs) 236
single-molecule real-time (SMRT) 9, 12
single-molecule real-time (SMRT)
sequencing 12
small nuclear ribonucleoproteins (snRNPS) 102
Spearman's rank correlation 184
spliced aligners
HISAT2 47–50
STAR 50–53
TopHat 43–47
spliced transcripts aligned to a reference
(STAR)
STAR 50–53
STAR-Fusion 145–149
STARsolo 242–247
spliceosome 101–103
splicing regulatory elements (SREs) 105
split read 137–144
SPLiT-Seq 239
Stringtie 61–63
subread 40–43

T-distributed stochastic embedding (t-SNE)
plot 252, 256
TopHat 43–47
toxicology 172–174
toxicogenomics
advantages and limitations 175–176
co-expression networks 178–182
databases 188–199
identification of DEGs 176–178
introduction 175
gene networks 182–187
signature matching 182–184
traditional medicine
databases 206–209
development of molecular markers 219–220

discovery of developmental mechanisms
218–219
discovery of secondary metabolites
214–218
drug repurposing 226
identification of molecular mechanisms
220–222
introduction 205–206
mining functional genes 210–213
Synergism 222–223
toxicity 223–225
transcriptome 2–3
transcripts per million (TPM) 68
transfer RNA (tRNA) 2
trans-splicing 132
Trimmomatic 29–30
Trinity 55–57

UCSC genome browser 84–85
unique molecular identifier (UMI) 236–237
unspliced aligners
Bowtie 36–40
seed-based methods 40–43

violin plot 252
visualization
Database for Annotation, Visualization,
and Integrated Discovery (DAVID)
91–95
heatmaps 86–90
Integrative Genomics Viewer (IGV)
81–84
UCSC Genome Browser 84–85
volcano plots 90–92
volcano plots 90–92

Weighted Gene Co-expression Network
Analysis (WGCNA) 185–187

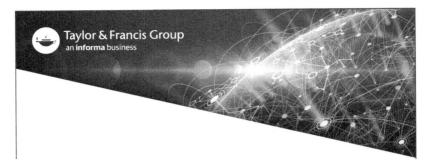